Charge Pump IC Design

Feng Pan

Mc Graw Hill Education

New York Chicago San Francisco
Athens London Madrid
Mexico City Milan New Delhi
Singapore Sydney Toronto

McGraw-Hill Education books are available at special quantity discounts to use as premiums and sales promotions or for use in corporate training programs. To contact a representative, please visit the Contact Us page at www.mhprofessional.com.

Charge Pump IC Design

1 2 3 4 5 6 7 8 9 0 QVS/QVS 20 19 18 17 16 15

ISBN 978-0-07-183677-7
MHID 0-07-183677-2

This book is printed on acid-free paper.

Sponsoring Editor
Mike McCabe

Editorial Supervisor
Donna M. Martone

Production Supervisor
Lynn M. Messina

Acquisitions Coordinator
Molly T. Wyand

Project Manager
Tanya Punj,
Cenveo® Publisher Services

Copy Editor
Patti Scott

Proofreader
Eina Malik

Indexer
Edwin Durbin

Art Director, Cover
Jeff Weeks

Composition
Cenveo Publisher Services

Contents

About the Author

Feng Pan received his BS degree in EECS from the University of California, Berkeley, and MS in Electrical Engineering from Stanford University. He is a Stanford Certified Program Manager (SCPM), and is currently working as a director of NSG at Micron Technology, Inc. on 3D VNAND technology. Mr. Pan previously held various senior management and leadership positions at SanDisk and AMD. Mr. Pan is coauthor of *Charge Pump Circuit Design*, and has 51 granted patents related to power management, LDO, charge pump architectures, charge pump regulations and applications, ADC design, op-amp designs, and NAND/NOR flash memory designs.

Preface

Over the last decade or so, the portable devices, hand-held devices, smart phones, SSDs, etc., have proliferated in our daily lives. Power management and power harvesting have become important research themes in analog designs to reduce power consumption and to use alternative and renewable energy sources instead of traditional fossil fuels. The current power management has branched off to become an emerging field of its own in analog design. For on-chip DC–DC converter, like charge pump, there have been a lot of new studies and researches since *Charge Pump Circuit Design* was first published in 2006. It is the time to reflect and have a new book to systematically discuss the theories, practical designs, and applications on charge pump from real practices to emphasize on performance, power efficiency, and area efficiency.

Charge Pump IC Design has the following key focuses to make it an advanced reference book for power management design professionals.

First, it introduces and compares the various power management and power conditioning circuits at system level and at chip level. It gives a systematic approach to derive the charge pump characteristics.

Second, it discusses a systematic approach on how to design charge pump from a new perspective, with the local optimization on basic charge pump stage, which leads to the optimum final design.

Third, it discusses various design approaches to address charge pump-related circuits and regulation schemes with key emphasis on performance power and area efficiency using real design examples. This allows readers to see practical approaches and trade-offs in actual design.

Fourth, it explains in details the negative feedback control and compensation in charge pump stability analysis with practical examples.

Fifth, it introduces the new concepts of using charge pump to design current-based ADC (analog-to-digital converter). This leads to brand new approaches for signal processing, power management control, etc.

This book is based upon many years of research works in charge pump and power management design in practice. It can be served as a reference book for analog and power management IC design engineers in industry, or graduate students in the fields of power management, power harvesting, or power electronics IC designs.

Feng Pan

Acknowledgments

Since publishing *Charge Pump Circuit Design* in 2006, it is now the right time to reflect and consolidate all the thoughts in a new book from the continuous learning and practicing of analog design in power management. This is a one-and-half-year project spanning many weekends and nights in San Francisco and Fremont while accompanying my daughter to go through her talent training. It is such a magical feeling to see her growth and improvement while I am able to wrap up this book at the same time with her graduation from the training.

This book has benefited from the works and studies of doing NAND flash memory design at SanDisk. The practices of circuit design, layout, and silicon debugging have helped me to grow more confident in analog design. Many thanks go to my colleagues at SanDisk for their fruitful discussions on various analog design topics. I also thank many of my colleagues at Micron for suggesting that I write this book. I am particularly grateful to Ramin Ghodsi's support from Micron management in writing this book.

My wife, Danae Deng, has encouraged me in the last sixteen years to set higher goals and to take on greater challenges in my life. She helps me to grow in many areas. I would not have been able to write this book without her full support. My beautiful daughter, Tiffany Pan, was born in 2006 when my first book was published. She has been a treasure to me and brings a lot of happiness to my life. It is with her encouragement that I am able to complete this project in time. She told me that she wants to write a book by herself someday like her dad. I am grateful to my parents Nianen Wang, Baohua Pan, and my brother Dean Pan for all their advice and support in my life.

The timely production of this book was made possible by the hard work of the staff at McGraw-Hill, especially, Donna M. Martone, Lynn M. Messina, Molly T. Wyand, and Jeff Weeks. I would like to thank Cenveo Publisher Services, particularly Tanya Punj and her team for their figure, proofreading (Eina Malik), grammar suggestions (Patti Scott), and indexing (Edwin Durbin). I am also pleased to express my gratitude to Mike McCabe at McGraw-Hill for initiating this project and helping me in many ways to complete this book.

I learned analog design from Stanford University and U.C. Berkeley. I want to express my gratitude to my schools and professors here.

Feng Pan
January 2015

CHAPTER 1

Regulators and Power Converters

In the semiconductor field, at either the system level or the chip level, the available power supplies given may not be suitable to meet all the power specifications of the loading chips or the loading circuits. The needs for power conversion between supplies or power supply conditioning may arise from one or more of the following design requirements: converting power between AC power supply and DC power supply, inverting the polarity of the power supply from the available power supply, modifying the magnitude of the power supply, improving the power efficiency of the power conversion or power conditioning circuits or reducing the output noise seen by loads from the given power supply. In general, the achievement of both high power efficiency and low output supply noise is the most challenging design goal in power converter or power conditioning design. With a single design solution it is difficult to achieve both design goals at the same time. Several designs solutions at different levels of integration are required to work together to achieve the highest power efficiency and lowest output supply noise in the overall system.

Based upon the location or the purpose of the solution, power circuit designs could be sorted into three main categories: (1) linear regulator, (2) switch mode power supply (SMPS), and (3) charge pump. A combination of SMPS and a low dropout (LDO), a combination of SMPS and a charge pump, or some other forms of combinations are common in today's overall system design. Each serves an important role in many electronic systems to ensure the overall power supply integrity of designs and power efficiency of the system as a whole.

1.1 Linear Regulator

A linear regulator is used in an electronic system to maintain a clean and steady power supply to the load. Despite variations or noises from the input power supply, or the faster switching nature of the loadings, a linear regulator should be able to (1) reject or filter noises from either the input supply or loadings and (2) maintain the target regulation within design specifications under all conditions. Several configurations of linear regulators are commonly used by designers. The high dropout (HDO) regulator and low dropout (LDO) regulator are two of the most popular architectures that could be found in many integrated circuits (ICs) and systems. Given the demands for higher power efficiency and higher power integrity with the proliferation of handheld devices storage and processing requirements in the cloud, the LDO regulator has gained significant popularity in recent years and has been spun off into a separate branch of study of its own, as have many other analog systems, such as data converter design and high-speed I/O design.

1.1.1 Input Power Supply Noise Rejection

Figure 1.1 illustrates the waveforms of the input power supply noise rejection of an ideal linear regulator with constant current load. With constant current load, for an ideal linear regulator with infinite bandwidth and gain, the regulator should be able to reject input supply noises from V_{cc} at any magnitude and at any frequency. The output supply V_{sup} would be regulated at the target value regardless of how the input supply V_{cc} varies.

Line regulation (LNR)[1-3] is an important DC performance parameter of the linear regulator. It refers to how much the output voltage changes with respect to the input power supply DC change. The input supply is at DC. It is equivalent to the DC gain

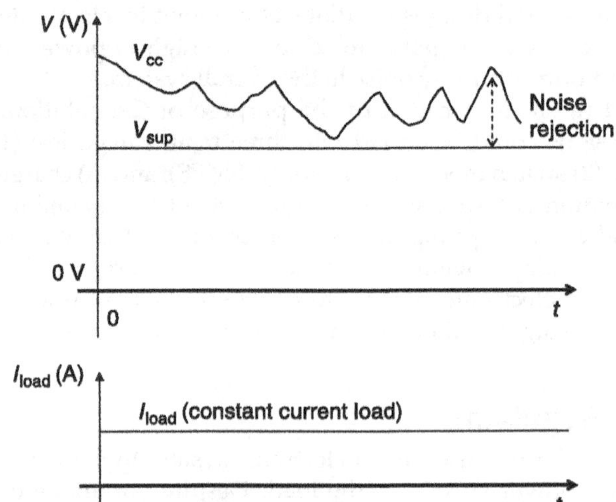

FIGURE 1.1 Input power supply noise rejection of ideal linear regulator with constant current load.

from the input power supply V_{cc} to the output power supply V_{sup}, as in Eq. (1.1).

$$\mathrm{LNR} = A_v(s=0) = \frac{\Delta V_{sup}}{\Delta V_{cc}} \tag{1.1}$$

$$\mathrm{PSR} \approx \frac{1}{A_v(s)} = \frac{\Delta V_{cc}(s)}{\Delta V_{sup}(s)} \tag{1.2}$$

The power supply rejection (PSR)[2] or ripple rejection, on the other hand, is the parameter used to measure the output supply rejection to the input supply variation over the entire frequency spectrum, as defined in Eq. (1.2). With limited bandwidth, linear regulator would not be able to reject input supply noise exceeds its closed-loop bandwidth. The input supply noise that exceeds the closed-loop bandwidth must be filtered out by the decoupling capacitance on the output. If input supply disturb causes a large signal excursion, regulator would not be able to response to hold the design specification. With increased bandwidth of circuits or added larger decoupling capacitance could address the issue.

1.1.2 Load Noise Rejection

Figure 1.2 illustrates the waveforms of the load noise rejection of the ideal linear regulator with constant input supply. With the input power supply V_{cc} unchanged, regardless of how the current load I_{load} switches, the ideal regulator with infinite bandwidth and gain should respond instantaneously to the change of load current with any

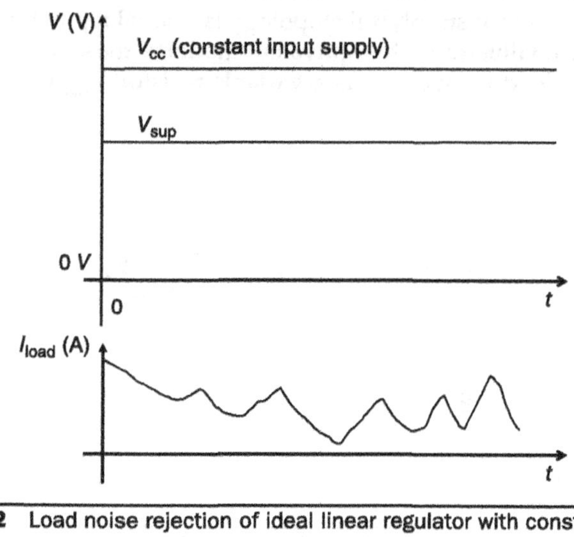

Figure 1.2 Load noise rejection of ideal linear regulator with constant input supply.

frequency and with any magnitude, and still be able to maintain the target output regulation voltage within design specifications.

Load regulation (LDR)[2] is an important DC performance parameter of a linear regulator. It refers to how much the output supply V_{sup} changes with respect to the change in load current I_{load}. The load current is DC. Load regulation is equivalent to the output impedance of the regulator at DC, as shown in Eq. (1.3).

$$\mathrm{LDR} = Z_{out}(s = 0) = \frac{\Delta V_{sup}}{\Delta I_{load}} \tag{1.3}$$

With limited bandwidth and gain of circuits, the linear regulator would not be able to reject load noise that exceeds its closed-loop bandwidth. The noise of the closed-loop bandwidth must be filtered out by the decoupling capacitance on the output. In a similar argument, if the load dump disturbance causes a very large signal excursion in the output of the regulator, the regulator may not be able to hold the design specification. Either adding more decoupling capacitance or increasing the bandwidth of regulator could address the issue.

1.1.3 Topologies of Linear Regulators

Linear regulators could be categorized into two basic configurations: serial regulators and shunt regulators. The topology of the serial or shunt regulator depends on whether the regulator is connected in parallel or series with the load.

1.1.3.1 Serial Regulator

If the regulator is in serial with the load and is in between the input supply and output supply, this topology is a serial regulator.

Figure 1.3 illustrates the equivalent model of the serial linear regulator. The load is modeled as a variable resistor R_{load} to mimic the

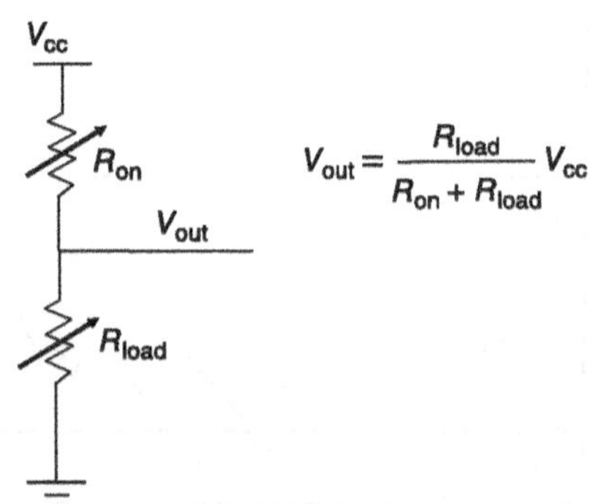

FIGURE 1.3 Equivalent model of serial linear regulator.

dynamic nature of the load switching. The serial regulator is modeled as a second variable resistor R_{on}, which is adjusted by the negative feedback control loop. The output voltage is given by Eq. (1.4).

$$V_{out} = \frac{R_{load}}{R_{on} + R_{load}} V_{cc} \tag{1.4}$$

Assuming the input supply V_{cc} was constant, as the load current varies, R_{load} would vary. To maintain the output voltage V_{out} unchanged, R_{on} would be changed by the regulation loop to ensure that the resistor divider ratio remains unchanged. This is the basic mechanism of how the serial regulator operates. Similarly, analysis can be applied to LNR. Assuming the output load current was constant, as the input supply V_{cc} varies, R_{on} would be changed by the regulation loop to maintain the target regulation voltage on output.

Figure 1.4 illustrates the circuit design model of the serial linear regulator. Negative feedback control is used in the design to control the regulation. Regardless of the causes of variations or changes, the variations on V_{out} are sampled and compared with a known reference. The negative feedback control loop would adjust R_{on} to bring V_{out} back on target.

1.1.3.2 Shunt Regulator

If the regulator is in parallel with the load in operation, this configuration is a shunt regulator.

Figure 1.5 illustrates the equivalent model of a shunt linear regulator. The load is modeled as a variable resistor R_{load} to mimic the dynamic nature of the load switching. The shunt regulator is modeled

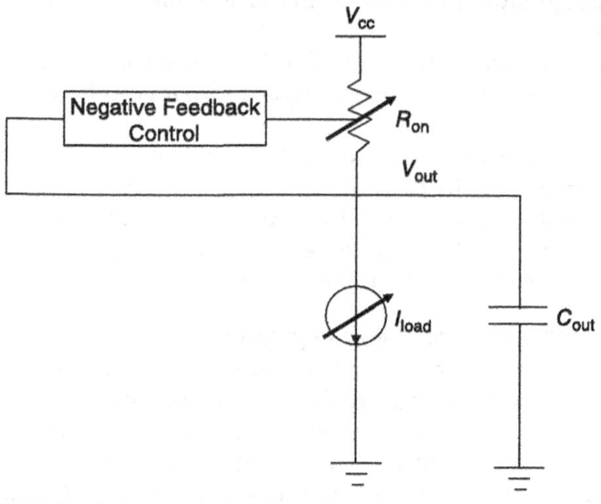

FIGURE 1.4 Serial linear regulator model.

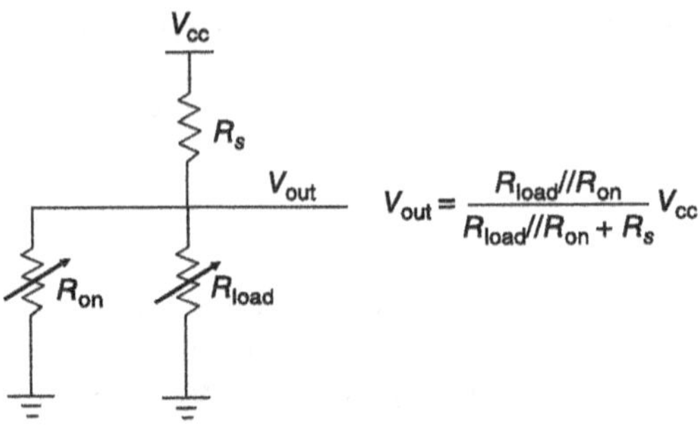

Figure 1.5 Equivalent model of shunt linear regulator.

as a second variable resistor R_{on} in parallel with R_{load} in between V_{out} and ground. And R_s is the impedance of power passing element connected between V_{cc} and V_{out}. Output voltage V_{out} could be derived from Eq. (1.5).

$$V_{out} = \frac{R_{load}//R_{on}}{R_{load}//R_{on} + R_s} V_{cc} \tag{1.5}$$

Assuming V_{cc} was constant for LDR, if the output load current changes, R_{load} would change. With negative feedback control, R_{on} would be modified accordingly to ensure the resistor divider ratio is unchanged. Shunting unneeded supply current away from the load is the basic mechanism of how shunt regulator operates. Since shunt regulator has to divert unused supply current to ground at any given time, it makes the shunt regulator much less efficient than the serial regulator, particularly under low-load or no-load current conditions.

The analysis for LNR is similar. With fixed load current, as V_{cc} varies, R_{on} would change accordingly to maintain the regulation in control.

Figure 1.6 illustrates the circuit design model for shunt linear regulator. Negative feedback control is applied in the design to control the value of R_{on}. Regardless of the causes of variation or noise on the output voltage V_{out}, the negative feedback control would adjust R_{on} so that V_{out} will be at target regulation level.

1.1.3.3 Application of Linear Regulators

A linear regulator is an element commonly used in many designs. It is simple and easy to design without the involvement of electromagnetic parts—such as inductors or transformers. Linear regulators are preferred if target output voltage and input supply voltage are not far apart. For heavy load operation, the SMPS can operate much more efficiently in power conversion. For applications where the target

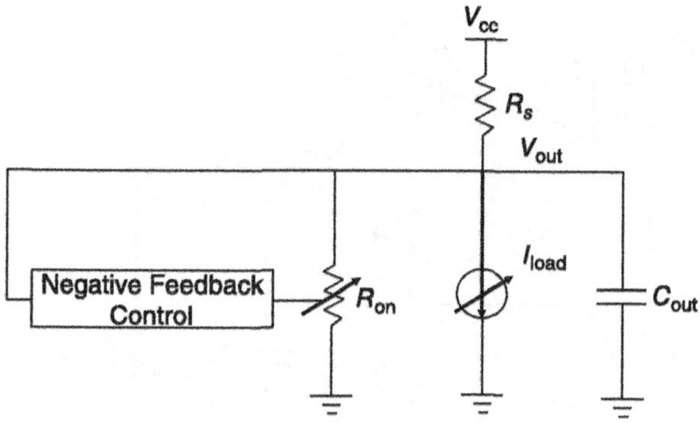

FIGURE 1.6 Shunt linear regulator model.

voltage and input supply differ to a greater degree, the linear regulator would not be efficient in the design.

$$\text{Power loss} = (V_{cc} - V_{out})I_{load} \tag{1.6}$$

$$\text{Power efficiency} = \frac{V_{out}I_{load}}{V_{cc}(I_{load} + I_{quiescent})} \approx \frac{V_{out}}{V_{cc}} \tag{1.7}$$

Power loss across the passing element in the serial regulator from Eq. (1.6) would be converted directly into heat. The maximum power conversion efficiency of a linear regulator is given by Eq. (1.7) if $I_{quiescent}$ is negligible compared with I_{load}. The limitation of adopting linear regulators in applications is the low power efficiency if the target voltage deviates significantly from the input supply, and significant power is required by load.

1.1.4 High Dropout Regulator

A high dropout (HDO) regulator is one special type of linear regulator, and it has been used for power conditioning in many areas because of the simplicity of its design.

1.1.4.1 Topology

Figure 1.7 illustrates the basic configuration of an HDO regulator. The passing element of the regulator is an N-channel MOSFET. If the error amplifier is supplied with the same input supply V_{cc}, that provides load current the maximum output voltage V_{sup} could be regulated as shown in Eq. (1.8).

$$V_{sup}(\max) = V_{cc} - V_{ov(P)} - V_{T(N_1)} - V_{ov(N_1)} \tag{1.8}$$

Here $V_{ov(P)}$ is the minimum overdrive voltage of the P-channel MOSFET in the output stage of the error amplifier. Also $V_{T(N_1)}$ and $V_{ov(N_1)}$ are the

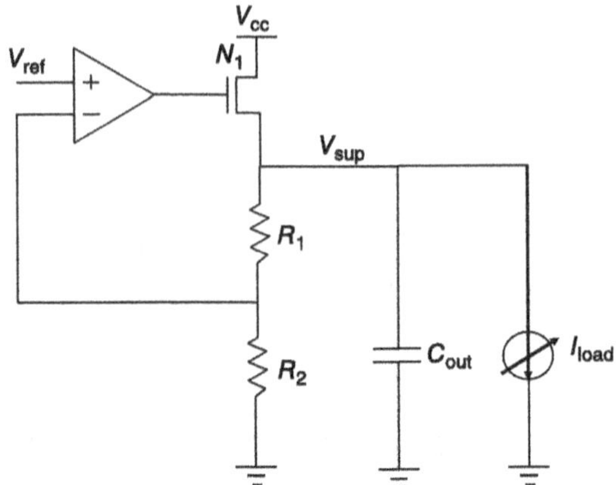

FIGURE 1.7 High dropout regulator.

associated threshold voltage with the body bias of N_1 and the overdrive voltage of N_1.

1.1.4.2 Advantages and Disadvantages

With moderate capacitance on the output supply V_{sup}, the HDO regulator could act as a single-POLE system, which is unconditionally stable. The dominant pole is located at the output node of the error amplifier. With the high g_m of the passing transistor N_1, the output node is a nondominant pole of the negative feedback system. Compensation of the HDO regulator is relatively simple, and the bandwidth of the closed-loop system is reasonably large. The PSR or power supply rejection ratio (PSRR) from the input supply is significantly improved with the N-channel MOSFET as the passing element. The PSR of the HDO regulator is eventually limited by the PSR of the error amplifier in the loop. These advantages are the main reasons why the HDO regulator is preferred over other architectures in design.

The disadvantage of the HDO regulator is mainly related to the higher headroom requirement and lower power conversion efficiency. Equation (1.8) illustrates the maximum output voltage level that could be achieved if the system had only a single power supply V_{cc}. The body bias of NMOS N_1 would increase with the increase of V_{sup}, as shown in Eq. (1.9). The overdrive voltage $V_{ov(N_1)}$ with respect to maximum load current $I_{load\ max}$ given in Eq. (1.10) determines the extra voltage headroom margin of design.

$$V_{T(N_1)} = V_{T_0(N_1)} + \gamma\left(\sqrt{V_{reg} + |2\theta_F|} - \sqrt{|2\theta_F|}\right) \tag{1.9}$$

$$V_{ov(N_1)} = \sqrt{\frac{2I_{load\ max}}{\mu_n C_{ox}(W/L)_{(N_1)}}} \tag{1.10}$$

The power loss in the HDO regulator due to passing element N_1 is given by Eq. (1.11). The first term is the minimum power consumption on N_1 in the HDO regulator, and the second term is simply the power consumption across N_1 depending on output regulation level and input supply voltage. From the design, it is obvious that power efficiency for the HDO regulator would not be high. The difference between V_{cc} and V_{sup} should be reduced. A lower threshold NMOS transistor should be used, and minimizing the overdrive voltage for both NMOS and PMOS would help to improve the HDO regulator's power efficiency.

$$P_{loss} = \text{maximum}[(V_{ov(P)} + V_{T(N_1)} + V_{ov(N_1)})I_{load}, (V_{cc} - V_{sup})I_{load})] \tag{1.11}$$

$$V_{high}(\text{min}) = V_{sup}(\text{max}) + V_{ov(P)} + V_{T(N_1)} + V_{ov(N_1)} \tag{1.12}$$

An alternative solution to improve the HDO regulator's power efficiency is to use a separate higher power supply V_{high} for the error amplifier as illustrated in Fig. 1.8. A small charge pump could generate V_{high} with its minimum output level defined by Eq. (1.12). The gain in power efficiency and headroom from passing element N_1 would dominate the power consumption from the small charge pump.

1.1.5 Low Dropout Regulator

The low dropout (LDO) regulator[1-2] is another special linear regulator with a target regulation level very close to the level of the input supply V_{cc}. With very high power efficiency, the LDO regulator has been used extensively in integrated circuit (IC) designs for power conditioning.

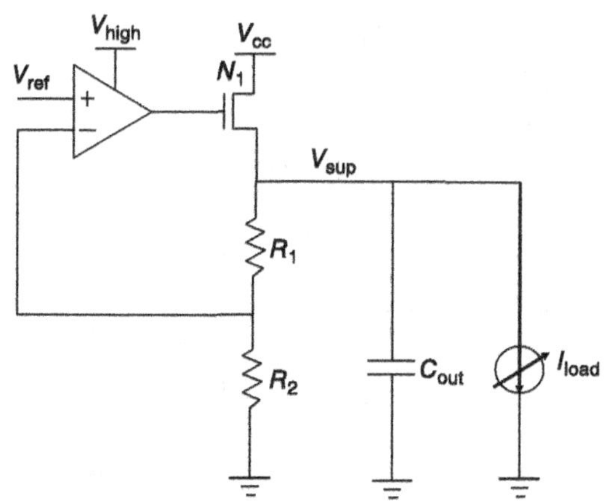

FIGURE 1.8 HDO regulator with additional high-power supply.

1.1.5.1 Topology

Figure 1.9 illustrates the basic topology of the LDO regulator. The passing element is a P-channel MOSFET, which is different from an N-channel MOSFET used in the HDO regulator. To pass high voltage, the gate of the PMOS does not need to be driven higher than the input supply V_{cc}. PMOS has higher current sourcing capability than NMOS.

$$V_{sup}(\max) = V_{cc} - V_{ov\ max(P_1)} \tag{1.13}$$

$$V_{ov\ max(P_1)} = \sqrt{\frac{2I_{load\ max}}{\mu_p C_{ox}(W/L)_{(P_1)}}} \tag{1.14}$$

With the LDO regulator, the maximum output voltage that V_{sup} could reach before the PMOS transistor enters the dropout region is given by Eq. (1.13). The maximum overdrive $V_{ov\ max(P1)}$ in Eq. (1.14) is a function of the maximum load current and device dimensions assuming technology is fixed. If the LDO regulator entered into the dropout region, there would not be enough loop gain left to correct the feedback error.

1.1.5.2 Advantages and Disadvantages

The advantage of the LDO regulator is obvious in terms of higher power efficiency. As long as the target output voltage V_{sup} is close to the input power supply V_{cc} before entering the dropout region, the power loss Eq. (1.15), on passing element P_1 is minimized by keeping $V_{ov\ max(P1)}$ small.

$$P_{loss} = \text{maximum}(V_{ov\ max(P_1)} I_{load}, (V_{cc} - V_{sup}) I_{load}) \tag{1.15}$$

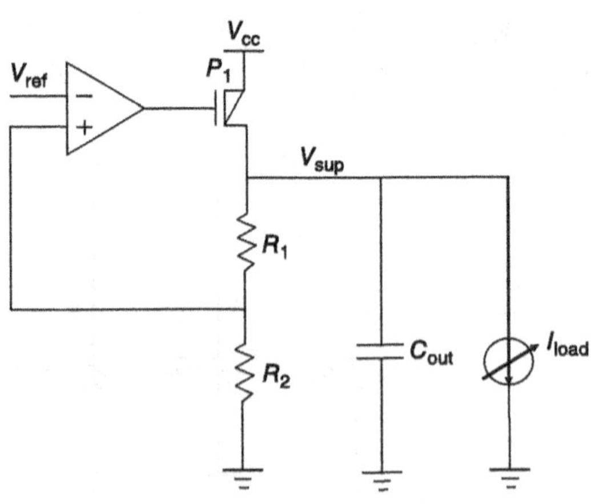

FIGURE 1.9 Low dropout regulator.

The difficulty of designing an LDO regulator lies in how to compensate the LDO for stability while still meeting all other design specifications. The load current could vary by a few orders. The power consumption budge could be limited. Internal compensation/external compensation, LDR, LNR, load dump response, PSRR, die size, and cost could require significant efforts in planning and actual design. This makes the design of the LDO regulator much more challenging than that of the HDO regulator.

1.2 Switch Mode Power Supply

The switch mode power supply (SMPS)[3] is a discrete power converter solution used widely at the system level to achieve very high power conversion efficiency in high-output power range. The output current ranges from 50 to approximately 100 mA to a few amperes, and the power conversion efficiency is in the range of 80% to 90%, or could be even higher.

The SMPS relies on electromagnetic components; an inductor or a transformer to store and convert energy. Energy converts between electrical and magnetic forms in alternating clock phases, and the power loss incurred between conversions is very little. With other types of power converters or power conditioning circuits, it is difficult to minimize the power loss on passing devices or parasitic components.

With the switching nature, the SMPS has relatively high output supply noise compared to other types of power converters. The SMPS is fully turned on to receive power from the input supply for a fraction of a clock period, and the passing element does not always conduct current between input and output. To minimize the output noise, the SMPS requires large decoupling capacitance on the output. During the on phase, the decoupling capacitance will absorb excess charge from the passing element that exceeds the load requirement. During the off phase, it will deplete the charge to the load while the passing element is turned off.

The SMPS minimizes the power loss on internal power switch. In the on phase, with the passing device fully turned on to charge the load, the resistance associated with the passing device would be minimized, and so it is the quiescent power loss on this device. In the off phase, with the passing device fully turned off, there is no quiescent power consumption on the passing device.

Based on the relationship between output supply and input supply, the SMPS could be grouped into a few categories: buck converter, boost converter, buck-boost converter, etc.

1.2.1 Buck Converter

The SMPS can generate output voltage lower than the input power supply, and this type of step-down only converter is called a buck converter[3–5]. Unlike the linear regulator, which converts any excess power to heat

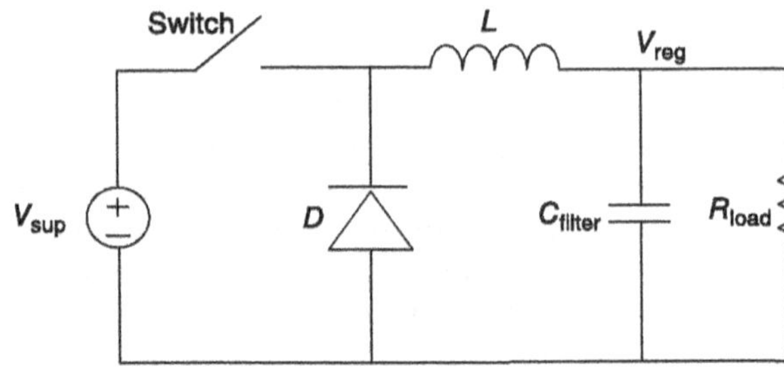

FIGURE 1.10 Buck converter configuration.

and wastes power, the buck converter can achieve high efficiency with power conversion between the electrical and magnetic forms without much loss. While stepping down the output voltage from the input supply, the output current capability is boosted up.

Figure 1.10 illustrates the basic circuit diagram of a buck converter.[3] The switch, diode, inductor, and filter capacitance form the basis of the converter. Here V_{sup} represents input DC supply, and the output load is simplified into a resistor R_{load}. The supporting circuits for regulation are not shown. The buck converter converts electrical energy from V_{sup} into magnetic energy stored in inductor L in the first clock phase. In the second clock phase, the magnetic energy is converted back from L to electrical energy and is delivered to the output. The conversion process is highly efficient, and higher efficiency allows the buck converter to handle the current load at much higher level than linear regulators can, in the range of a few hundred milliamperes to a few amperes.

The principle of a buck converter could be explained by using voltage–current relationship of inductor L when the switch is turned on or off.

1.2.1.1 On State

Figure 1.11 illustrates the on state of a buck converter after the switch is closed. Current flows in the directions outlined by the dashed arrows. Before the switch is closed, there is no current flow in the circuits. After the switch is closed, the inductor current will gradually increase over time. The polarity of the potential across inductor L opposes the change in inductor current. Over time, as the magnitude of the inductor current increases, the rate of current change would gradually decrease. The voltage across the inductor that opposes the change of inductor current will decrease as time elapses. Magnetic energy is gradually built up in inductor L. If given enough time, V_{reg} would approach V_{sup}. During the on state, there is no current flow through diode D since it is reverse-biased. Also C_{filter} would absorb any excess energy that R_{load} could not consume during the on state.

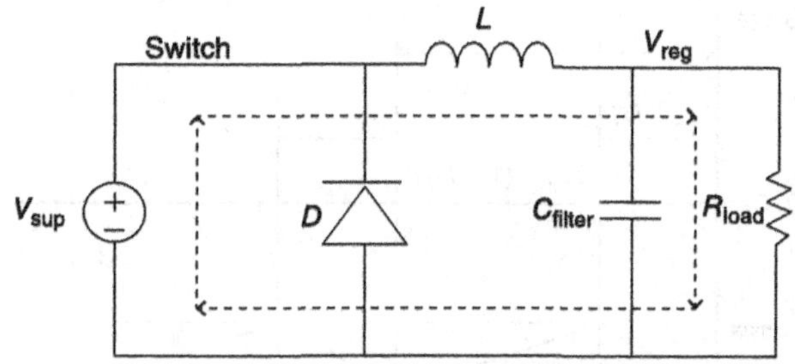

FIGURE 1.11 On state of buck converter.

1.2.1.2 Off State

Figure 1.12 illustrates the off state of a buck converter after the switch is opened. Current flows in the directions outlined by the dashed arrows. Before the switch is opened, a current flows in inductor *L*. After the switch is opened, diode *D* is turned on to close the current conduction loop. The inductor current would decrease over time from its peak value at the end of the on state. The polarity of voltage across the inductor would change and oppose the decrement of current through inductor *L*. If R_{load} consumes more current than the inductor could provide, then C_{filter} will discharge to provide the excess power needed by the load.

The buck converter has two modes of operation: continuous mode and discontinuous mode. It depends on whether the inductor current ever drops to zero during the conversion process.

1.2.1.3 Continuous Mode

If the inductor current never falls to zero at the end of the off state, then the buck converter is operating in continuous mode.

Figure 1.13 illustrates the current–voltage relationship of the inductor in continuous mode operation for a buck converter.

By assuming ideal circuit elements and an infinite C_{filter} to be used in design, it will be easy to derive the analysis and control scheme for

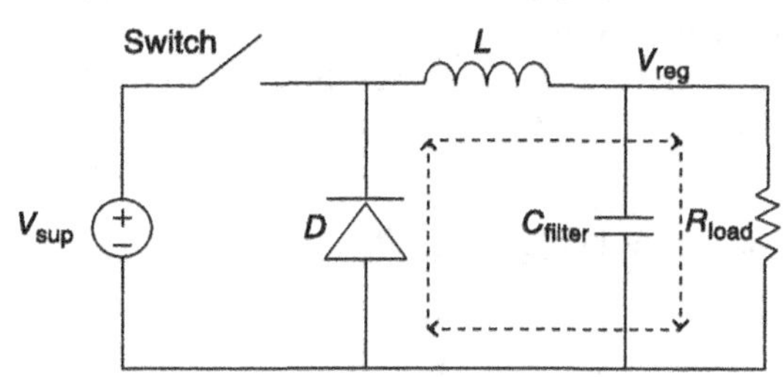

FIGURE 1.12 Off state of buck converter.

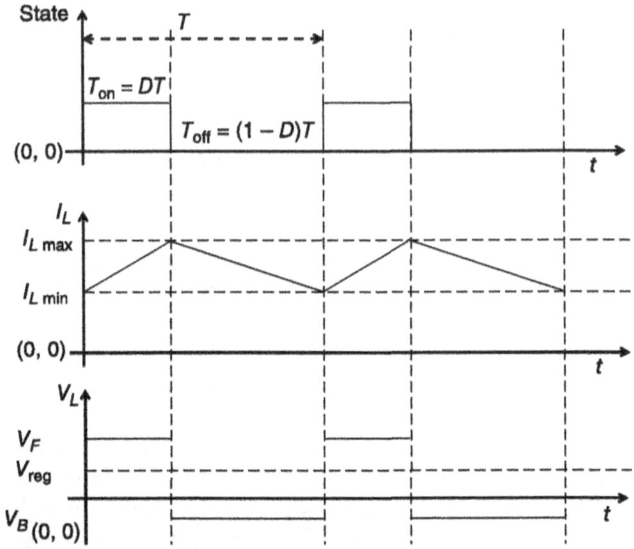

FIGURE 1.13 Continuous mode of a buck converter.

continuous mode. From the energy point of view, if the buck converter operated in steady state, then the energy added to inductor L during the on state and the energy dissipated from inductor L during the off state should be same. As shown in Eq. (1.16), there should not be an energy change in the inductor at the end of each cycle under the equilibrium condition. Equation (1.17) signals that the change of inductor current over one commutation cycle should be zero.

$$\frac{1}{2}L\Delta I_{L\,on}^2 - \frac{1}{2}L\Delta I_{L\,off}^2 = 0 \tag{1.16}$$

$$\Delta I_{L\,on} + \Delta I_{L\,off} = 0 \tag{1.17}$$

The analysis assumes ideal circuit elements. Here V_D is zero voltage for the diode, and there is infinite capacitance C_{filter} on the output. Voltage V_{reg} would not change over the commutation cycle for the analysis. During the on state, Eq. (1.18) gives the total change of inductor current at the end of the on state. Equation (1.19) gives the total change of inductor current at the end of the off state.

$$\Delta I_{L\,on} = \frac{1}{L}\int_0^{T_{on}} (V_{sup} - V_{reg})dt = \frac{T_{on}}{L}(V_{sup} - V_{reg}) \tag{1.18}$$

$$\Delta I_{L\,off} = \frac{1}{L}\int_{T_{on}}^{T} (0 - V_{reg})dt = -\frac{T_{off}}{L}V_{reg} \tag{1.19}$$

$$T_{on}(V_{sup} - V_{reg}) - T_{off}V_{reg} = 0 \quad (1.20)$$

$$\frac{V_{reg}}{V_{sup}} = D \quad (1.21)$$

With equilibrium conditions and proper substitutions, Eqs. (1.20) and (1.21) could be derived to show the ratio of V_{reg} to V_{sup}. Voltage V_{reg} is proportional to the input supply voltage V_{sup} as a function of D, which is on duty of cycle bounded between 0 and 1. Voltage V_{reg} could be 0 V if the switch is always open, or at its maximum if the switch is always closed to pass V_{sup} directly to V_{reg}.

In terms of regulation for a buck converter, the pulse width modulation (PWM) is applied to adjust the duty of cycle D to turn on or turn off the switch based upon the target regulation level and input supply V_{sup}.

1.2.1.4 Discontinuous Mode

If the inductor current ever falls to zero in the middle of or at the end of the off state, the buck converter is operating in discontinuous mode. Discontinuous mode occurs when the power consumption on the current load is too small.

Figure 1.14 illustrates that the current–voltage relationship of the inductor operates in discontinuous mode of the buck converter. By assuming ideal circuit elements and infinite C_{filter} to be used in design, it will be easy to derive the analysis and control scheme of

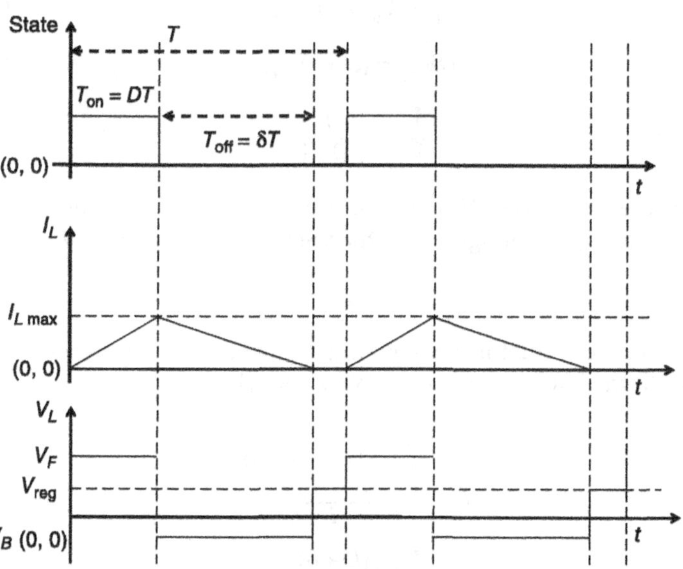

FIGURE 1.14 Discontinuous mode of a buck converter.

the discontinuous mode. From an energy point of view, if the buck converter is operating in steady state, then energy being added to inductor L during the on state and energy being dissipated from inductor L during the off state should be the same. As shown in Eq. (1.22), excess energy should not be added to the inductor at the end of each commutation cycle. Equation (1.23) indicates the total current change in inductor after one cycle of the on state and off state in discontinuous mode.

$$\frac{1}{2}L\Delta I_{L\,\mathrm{on}}^2 - \frac{1}{2}L\Delta I_{L\,\mathrm{off}}^2 = 0 \tag{1.22}$$

$$\Delta I_{L\,\mathrm{on}} + \Delta I_{L\,\mathrm{off}} = 0 \tag{1.23}$$

The analysis assumes ideal circuit elements. Voltage V_D is zero for the diode, and there is infinite capacitance C_{filter} on the output. Voltage V_{reg} would not change over the commutation cycle for the analysis. During the on state, Eq. (1.24) gives the total change of inductor current at the end of the on state. Equation (1.25) gives the total change of inductor current at the end of the off state.

$$\Delta I_{L\,\mathrm{on}} = \frac{1}{L}\int_0^{T_{\mathrm{on}}} (V_{\mathrm{sup}} - V_{\mathrm{reg}})\,dt = \frac{T_{\mathrm{on}}}{L}(V_{\mathrm{sup}} - V_{\mathrm{reg}}) \tag{1.24}$$

$$\Delta I_{L\,\mathrm{off}} = \frac{1}{L}\int_{T_{\mathrm{on}}}^{T} (0 - V_{\mathrm{reg}})\,dt = -\frac{\delta T}{L}V_{\mathrm{reg}} \tag{1.25}$$

$$T_{\mathrm{on}}(V_{\mathrm{sup}} - V_{\mathrm{reg}}) - T_{\mathrm{off}}V_{\mathrm{reg}} = 0 \tag{1.26}$$

$$DV_{\mathrm{sup}} = (D+\delta)V_{\mathrm{reg}} \tag{1.27}$$

$$\frac{V_{\mathrm{reg}}}{V_{\mathrm{sup}}} = \frac{D}{D+\delta} \tag{1.28}$$

Equations (1.26) to (1.28) could be derived to show the ratio of output voltage to input voltage. From the view of the current load angle,

$$\overline{I_L} = I_{\mathrm{load}} \tag{1.29}$$

The average inductor current over the commutation cycle must be equal to the load current as shown in Eq. (1.29).

$$\begin{aligned}\overline{I_L} &= \frac{\int_0^{T_{\mathrm{on}}} I_{\mathrm{on}}\,dt + \int_{T_{\mathrm{on}}}^{T_{\mathrm{off}}} I_{\mathrm{off}}\,dt}{T} \\ &= \frac{I_{L\,\mathrm{max}}(D+\delta)}{2} = I_{\mathrm{load}}\end{aligned} \tag{1.30}$$

Since $I_{L\max}$ could be derived as Eq. (1.31), Eq. (1.32) could be derived from Eq. (1.28):

$$I_{L\max} = \frac{V_{sup} - V_{reg}}{L} DT \tag{1.31}$$

$$\delta = \frac{V_{sup} - V_{reg}}{V_{reg}} D \tag{1.32}$$

Equation (1.30) could be rewritten as Eq. (1.33):

$$\begin{aligned} \overline{I_L} &= \frac{\left(\int_0^{T_{on}} I_{on}\, dt + \int_{T_{on}}^{T_{off}} I_{off}\, dt \right)}{T} \\ &= \frac{(V_{sup} - V_{reg}) DT(D+\delta)}{2L} = I_{load} \\ &= \frac{(V_{sup} - V_{reg}) DT[D + D(V_{sup} - V_{reg})/V_{reg}]}{2L} = I_{load} \end{aligned} \tag{1.33}$$

Solving the equation for V_{reg} would generate Eq. (1.34).

$$V_{reg} = V_{sup} \frac{1}{2LI_{load}/D^2 V_{sup} T + 1} \tag{1.34}$$

Compared with Eq. (1.21) of the continuous mode of a buck converter, the discontinuous mode output voltage is a much more complicated function of variables—input supply V_{sup}, load current I_{load}, and inductor L—on duty of cycle D and cycle period T. The nonlinear relationship in Eq. (1.34) means regulation control would be much more complicated. Pulse width modulation and pulse frequency modulation (PFM) could be applied in regulation control to maximize the power efficiency.

1.2.2 Boost Converter

The SMPS can generate output voltage higher than input supply. This type of SMPS is called a boost converter.[3-5] Since power conversion is stepping up the output voltage, the output current capability would reduce. The boost converter can achieve very high power efficiency through energy conversion between the electric field and magnetic field.

Figure 1.15 illustrates the basic circuit diagram of a boost converter. The switch, diode, inductor, and filter capacitance form the basis of the design. Voltage V_{sup} represents the input DC supply, and the output load is simplified into a resistor R_{load}. The supporting circuits for regulation are not shown. The concept of a boost converter is to convert electric energy from the input supply V_{sup} to magnetic energy stored in inductor L in the first phase. In the second phase, the magnetic energy

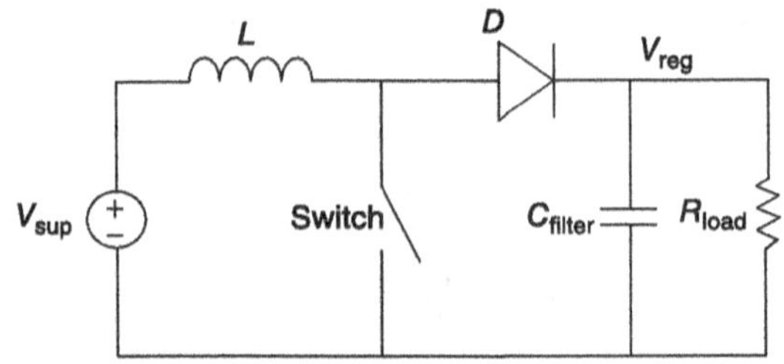

Figure 1.15 Boost converter configuration.

is converted back to electric energy and is delivered to the load. The difference between the boost converter and buck converter lies in simply rearranging the same circuit components in the configuration.

The principle of a buck converter can be explained using the voltage–current relationship of inductor L when the switch is turned on or off.

1.2.2.1 On State

Figure 1.16 illustrates the on state of a buck converter when the switch is closed. Current is flowing in the direction outlined by the dashed arrows. We assume that before the switch is closed, there is no current flow in the circuits. After the switch is closed, the inductor current would gradually increase over time. The polarity of potential across inductor L opposes the increment of inductor current. Over time, as the magnitude of inductor current increases, the rate of current change would gradually decrease. The voltage across inductor L opposing the change of inductor current will decrease. The magnetic energy is gradually built up in the inductor. During the on state, there is no current flow through the diode since it is reverse-biased. The C_{filter} will provide the energy needed by R_{load}.

1.2.2.2 Off State

Figure 1.17 illustrates the off state of a buck converter when the switch is opened. Current flows in the direction outlined by the dashed arrows. Before the switch is open, a maximum amount of current flows in

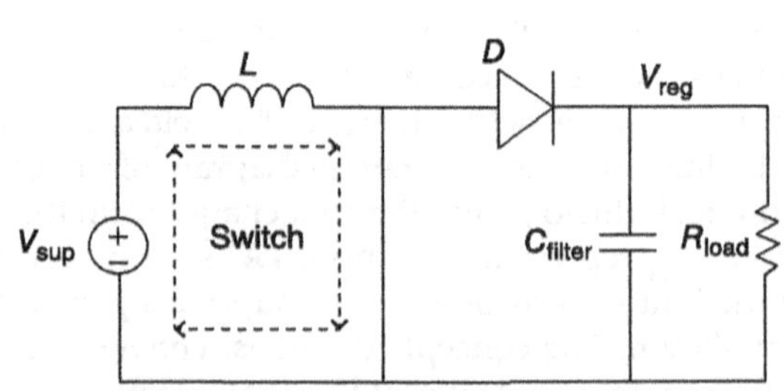

Figure 1.16 On state of a boost converter.

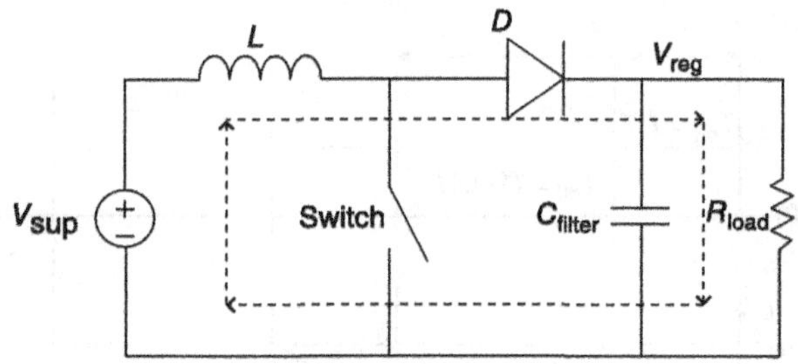

FIGURE 1.17 Off state of a boost converter.

inductor L. After the switch is opened, the inductor current must flow through the load R_{load}, and it will decrease over time. The polarity of voltage across the inductor will change and will add to V_{sup}, and will oppose the decrement of inductor current. In case the inductor provides higher current than R_{load} could consume, C_{filter} absorbs any excess energy from the inductor in the off state.

Similar to the operations of the buck converter, the boost converter has two distinctive modes of operation: continuous mode and discontinuous mode. It depends on whether the inductor current ever drops to zero during the conversion process.

1.2.2.3 Continuous Mode

If the inductor current never falls to zero at end of the off state, then the boost converter is operating in continuous mode. Figure 1.18 illustrates the current–voltage relationship of the inductor in continuous mode of the boost converter. Assume ideal circuit elements and an infinite C_{filter} are used in the design. Then it will be easy to derive the analysis and control scheme of continuous-mode operation. From the energy point of view, if the boost converter is operating in steady state, then the energy added to inductor L in the on state and the energy dissipated from inductor L in the off state should be same. As shown in Eq. (1.35), no excess energy should be injected or taken away from the inductor without breaking the equilibrium condition.

$$\frac{1}{2}\Delta I_{L\,on}^2 - \frac{1}{2}\Delta I_{L\,off}^2 = 0 \tag{1.35}$$

$$\Delta I_{L\,on} + \Delta I_{L\,off} = 0 \tag{1.36}$$

Equation (1.36) shows that the total change of inductor current over one commutation cycle should be zero.

$$\Delta I_{L\,on} = \frac{1}{L}\int_0^{T_{on}} (V_{sup} - 0)\,dt = \frac{T_{on}}{L} V_{sup} \tag{1.37}$$

$$\Delta I_{L\,off} = \frac{1}{L}\int_{T_{on}}^{T_{off}} (V_{sup} - V_{reg})\,dt = \frac{T_{off}}{L}(V_{sup} - V_{reg}) \tag{1.38}$$

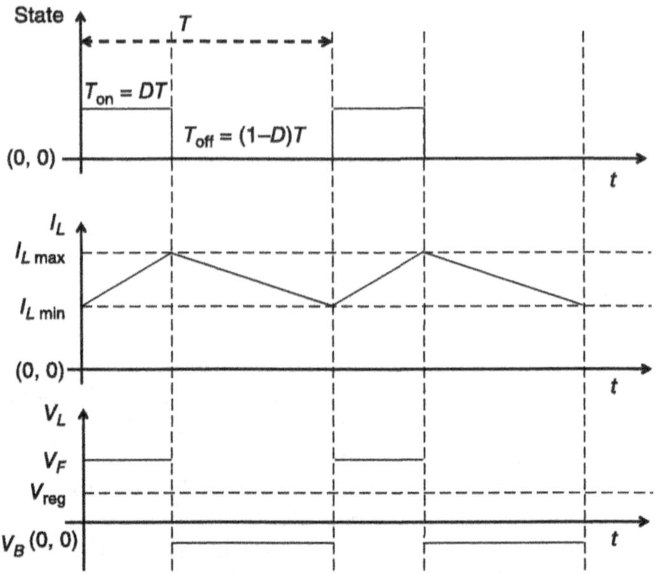

FIGURE 1.18 Continuous mode of a buck converter.

The analysis assumes ideal circuit elements. Voltage V_D is 0 for the diode, and there is infinite capacitance C_{filter} on the output. Voltage V_{reg} would not change over the commutation cycle for the analysis. During the on state, Eq. (1.37) gives the total change of inductor current at the end of on state. Equation (1.38) gives the total change of inductor current at the end of the off state.

$$T_{on}V_{sup} + T_{off}(V_{sup} - V_{reg}) = 0 \tag{1.39}$$

$$\frac{V_{reg}}{V_{sup}} = \frac{1}{1-D} \tag{1.40}$$

Due to equilibrium and with substitutions, Eqs. (1.39) and (1.40) could be derived to show that the ratio of output voltage V_{reg} to input voltage V_{sup} is a function of the on duty D. Here D is bounded between 0 and 1. Voltage V_{reg} could be equal to V_{sup} if $D = 0$, which means the switch is always open and input supply is connected directly with output supply. If D approached 1, then V_{reg} would approach infinity.

In terms of regulation for a boost converter, PWM is applied to adjust the duty D to turn on or off the switch based upon the target regulation level and input supply V_{sup}.

1.2.2.4 Discontinuous Mode

If the inductor current ever falls to zero in the middle of or at end of the off state, then the boost converter is operating in the discontinuous mode. Discontinuous mode occurs when the power consumption of the load is too small.

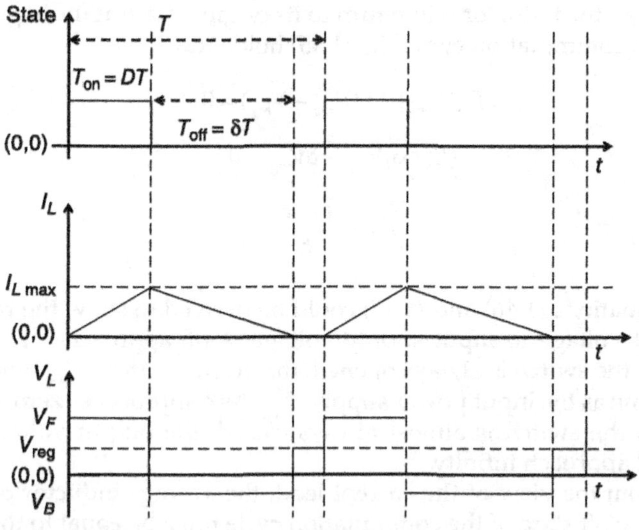

FIGURE 1.19 Discontinuous mode of a boost converter.

Figure 1.19 illustrates that the current–voltage relationship of the inductor operates in the discontinuous mode of the boost converter. If we assume ideal circuit elements and infinite C_{filter} are used in the design, it will be easy to derive the analysis and control scheme of the discontinuous mode. From the energy point of view, if the buck converter is operating in steady state, then the energy added to inductor L during the on state and the energy dissipated from inductor L during the off state should be same. As shown in Eq. (1.41), excess energy should not be injected or taken away from the inductor without breaking the equilibrium condition. Equation (1.42) indicates there is no net current change after one commutation cycle.

$$\frac{1}{2}L\Delta I_{L\,on}^2 - \frac{1}{2}L\Delta I_{L\,off}^2 = 0 \tag{1.41}$$

$$\Delta I_{L\,on} + \Delta I_{L\,off} = 0 \tag{1.42}$$

The analysis assumes ideal circuit elements. Voltage V_D is zero for the diode, and there is infinite capacitance C_{filter} on the output. Voltage V_{reg} would not change over the commutation cycle for the analysis. During the on state, Eq. (1.43) gives the total change of inductor current at the end of the on state. Equation (1.44) gives the total change of inductor current at the end of the off state.

$$\Delta I_{L\,on} = \frac{1}{L}\int_0^{T_{on}} (V_{sup} - 0)dt = \frac{T_{on}}{L}V_{sup} \tag{1.43}$$

$$\Delta I_{L\,off} = \frac{1}{L}\int_{T_{on}}^{T} (V_{sup} - V_{reg})dt = \frac{\delta T}{L}(V_{sup} - V_{reg}) \tag{1.44}$$

Since the inductor will return to its original state at the beginning of each commutation cycle, Eq. (1.45) holds true.

$$T_{on}V_{sup} + \delta T(V_{sup} - V_{reg}) = 0 \tag{1.45}$$

$$(D+\delta)V_{sup} - \delta V_{reg} = 0 \tag{1.46}$$

$$\frac{V_{reg}}{V_{sup}} = \frac{D+\delta}{\delta} \tag{1.47}$$

Equations (1.46) and (1.47) could be derived to show the ratio of output voltage to input supply voltage. As δ approaches 1, which means the switch is always opened, the output voltage V_{reg} would be the same as the input power supply V_{sup}. As δ approaches zero, which means the switch is almost always closed, the output voltage V_{reg} would approach infinity.

From the view of the current load, the average inductor current over the off state of the commutation cycle must be equal to the load current shown in Eq. (1.48) and Eq. (1.49).

$$\overline{I_L} = I_{load} \tag{1.48}$$

$$\begin{aligned}\overline{I_{L\,off}} &= \int_{T_{on}}^{T_{off}} \frac{I_{off}\,dt}{T}\\ &= \frac{I_{L\,max}\delta}{2} = I_{load}\end{aligned} \tag{1.49}$$

Since $I_{L\,max}$ could be derived as Eq. (1.50) and δ could be derived as Eq. (1.51) from Eq. (1.47),

$$I_{L\,max} = \frac{V_{sup}}{L}DT \tag{1.50}$$

$$\delta = \frac{V_{sup}}{(V_{reg} - V_{sup})}D \tag{1.51}$$

Equation (1.48) could be rewritten as Eq. (1.52).

$$\begin{aligned}\overline{I_L} &= \frac{V_{sup}DT}{2L}\frac{V_{sup}D}{V_{reg} - V_{sup}} = I_{load}\\ &= \frac{V_{sup}^2 D^2 T}{2L(V_{reg} - V_{sup})} = I_{load}\end{aligned} \tag{1.52}$$

Solving the equation for V_{reg} would generate Eq. (1.53):

$$\frac{V_{reg}}{V_{sup}} = 1 + \frac{V_{sup}D^2T}{2LI_{load}} \tag{1.53}$$

Compared with Eq. (1.40) of the continuous mode of the boost converter, in discontinuous mode the ratio between the output voltage V_{reg} and input supply voltage V_{sup} is a much more complicated function of many parameters: input supply V_{sup}, load current I_{load}, inductor L, on duty of clock D and cycle period T. The nonlinear relationship in Eq. (1.53) means the regulation control is much more complicated for boost converters operating in discontinuous mode. PWM and PFM could be applied in regulation control to maximize power efficiency.

1.2.3 Split-π Converter (Boost–Buck Converter)

A split-π converter[6] can generate output voltage either lower or higher than the input supply, as illustrated in Fig. 1.20. In topology it is equivalent to the cascading of a boost converter, then followed by a buck converter in series.

The boost converter generates output voltage higher than the input supply, and capacitance C_1 absorbs excess charge in voltage stepping up. The buck converter followed steps down the boosted voltage from the output preceding boost converter. Theoretically output voltage V_{reg} could be scaled up or down in any ratio from input supply V_{sup}. In theory with a modified configuration, if a boost converter follows a buck converter in series, the derived output voltage could be scaled either lower or higher than the input supply too.

1.2.4 Inverting Converter

The SMPS can generate output voltage in the opposite polarity of that of the input supply. Output amplitude could be scaled either lower or higher in any ratio to the input supply amplitude. This is achieved by a buck–boost configuration or inverting converter[3–5].

Figure 1.21 illustrates the basic topology of an inverting converter. Similar to the analysis of either the buck or boost converter, operations could be analyzed in terms of the voltage–current relationship of inductor L.

1.2.4.1 On State

Figure 1.22 illustrates the on state of the inverting converter when the switch is closed. Current flows in the direction outlined by the dashed arrows. Before the switch is closed, there is no current flow in the circuits.

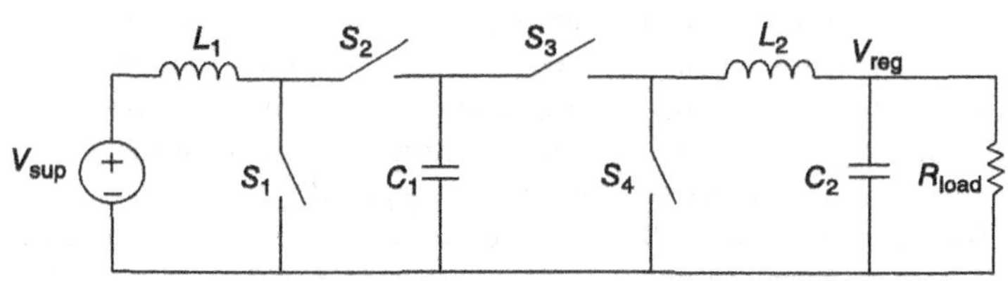

Figure 1.20 Split-π converter configuration.

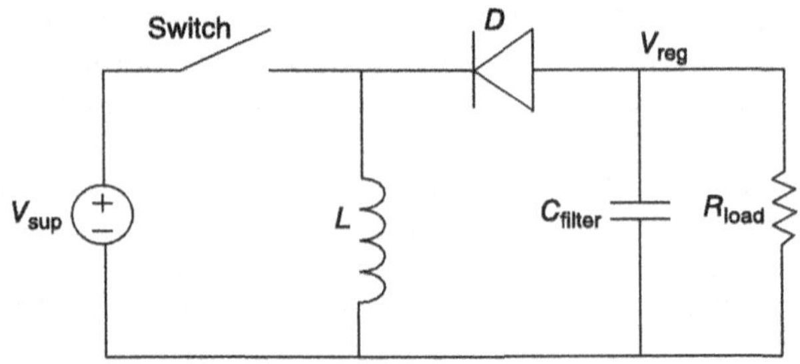

FIGURE 1.21 Inverting converter configuration.

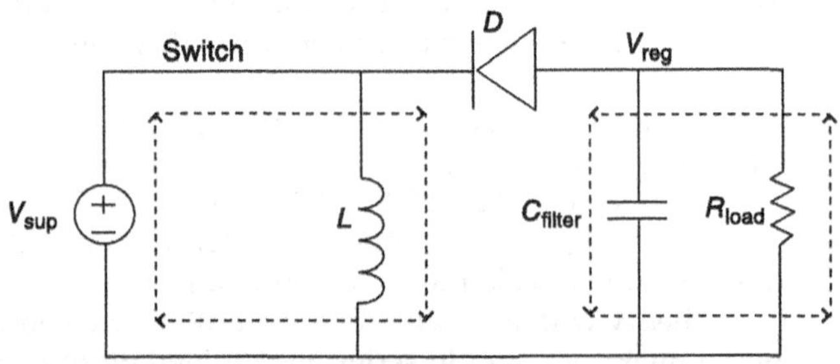

FIGURE 1.22 On state of inverting converter.

After the switch is closed, the inductor current will gradually increase over time. The polarity of the potential across inductor L opposes the increment of the inductor current. Over time, as the magnitude of the inductor current increases, the rate of current change will gradually decrease. The voltage across the inductor that opposes the inductor current change would decrease. The magnetic energy is gradually built up in the inductor. During the on state, there is no current flow through the diode since it is reverse-biased; C_{filter} will sustain the current flow of R_{load}.

1.2.4.2 Off State

Figure 1.23 illustrates the off state of an inverting converter when the switch is opened. Current flows in the direction outlined by the dashed arrows. Before the switch is open, a current flows in inductor L. After the switch is opened, the inductor current will decrease over time. The polarity of voltage across the inductor would switch, and oppose the decrement of inductor current. Since the top node of the inductor is turning negative, diode D is forward-biased and both R_{load} and C_{filter} are discharging to negative potential by the inductor current. Since V_{reg} is a more negative potential than the ground node, the polarity of V_{reg} is inverted from input supply V_{sup}.

The inverting converter has two modes of operation: continuous mode and discontinuous mode. It depends on whether the inductor current ever drops to zero during the conversion process.

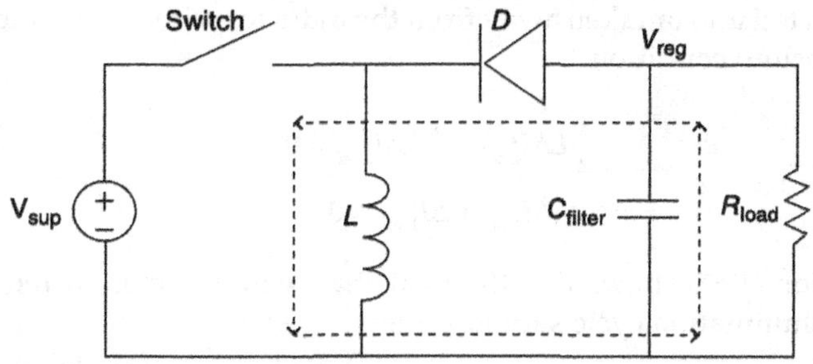

Figure 1.23 Off state of inverting converter.

1.2.4.3 Continuous Mode

If the inductor current never falls to zero at the end of the off state, the inverting converter operates in continuous mode. Figure 1.24 illustrates the current–voltage relationship of the inductor in continuous mode of the inverting converter. Assuming ideal circuit elements and infinite C_{filter} are used in the design, it will be easy to derive the analysis and control scheme of continuous mode.

From an energy point of view, if the inverting converter is operating in steady state, then the energy added to inductor L during the on state and the energy dissipated from inductor L during the off state should be the same. As shown in Eq. (1.54), excess energy should not

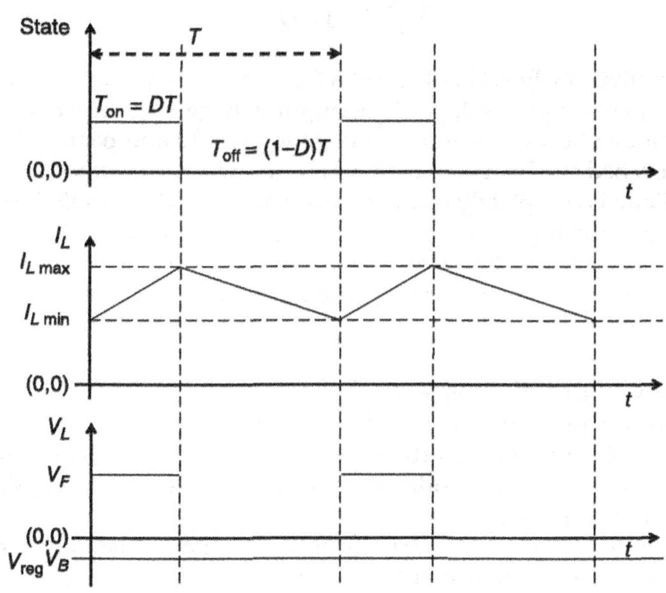

Figure 1.24 Continuous mode of an inverting converter.

be injected into or taken away from the inductor without breaking the equilibrium condition.

$$\frac{1}{2}L\Delta I^2_{L\,\mathrm{on}} - \frac{1}{2}L\Delta I^2_{L\,\mathrm{off}} = 0 \tag{1.54}$$

$$\Delta I_{L\,\mathrm{on}} + \Delta I_{L\,\mathrm{off}} = 0 \tag{1.55}$$

Equation (1.55) shows that the total change of inductor current over one commutation cycle should be zero.

$$\Delta I_{L\,\mathrm{on}} = \frac{1}{L}\int_0^{T_{on}} (V_{sup} - 0)\,dt = \frac{T_{on}}{L}V_{sup} \tag{1.56}$$

$$\Delta I_{L\,\mathrm{off}} = \frac{1}{L}\int_{T_{on}}^{T} (V_{reg} - 0)\,dt = \frac{T_{off}}{L}V_{reg} \tag{1.57}$$

The analysis assumes ideal circuit elements. Voltage V_D is zero for the diode, and there is infinite capacitance C_{filter} on the output. Voltage V_{reg} would not change over the commutation cycle for the analysis. During the on state, Eq. (1.56) gives the total change of inductor current at the end of the on state. Equation (1.57) gives the total change of inductor current at the end of the off state.

$$T_{on}V_{sup} + T_{off}V_{reg} = 0 \tag{1.58}$$

$$\frac{V_{reg}}{V_{sup}} = -\frac{D}{1-D} \tag{1.59}$$

At equilibrium, Eqs. (1.58) and (1.59) could be derived to show that in the ratio of output voltage V_{reg} to input voltage V_{sup}, duty of cycle. D is bounded between 0 and 1. Voltage V_{reg} would approach 0 V if D is approached 0. This is the case when the switch is always open and the diode is completely off. The potential across R_{load} would be 0 V. If D is approaching 1, then V_{reg} would approach negative infinity.

In terms of regulation for a buck converter, operates in continuous mode PWM is applied to adjust the duty cycle D with negative feedback control.

1.2.4.4 Discontinuous Mode

If the inductor current ever falls to zero in the middle of or near end of the off state, the inverting converter operates in discontinuous mode. Discontinuous mode occurs when the power consumption on the load is very small.

Figure 1.25 illustrates that the current–voltage relationship of the inductor operates in discontinuous mode of the inverting converter. If we assume ideal circuit elements and infinite C_{filter} are used in the design, it will be easy to derive the analysis and control scheme of

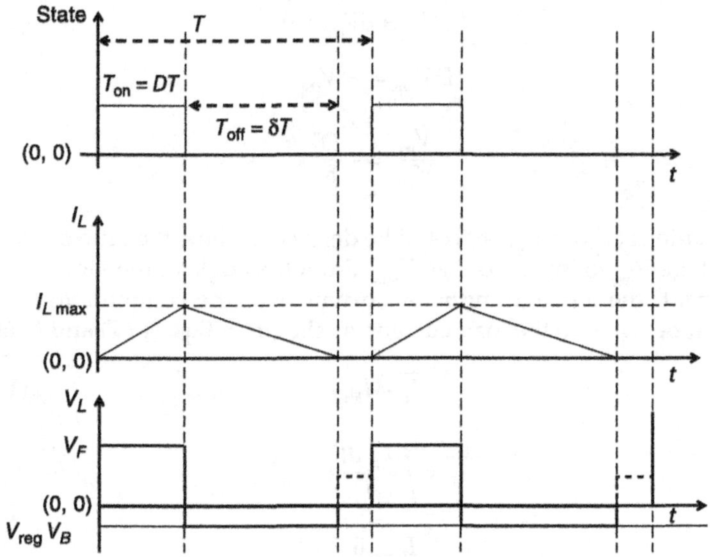

FIGURE 1.25 Discontinuous mode of a buck converter.

the discontinuous mode. From the energy point of view, if the inverting converter was operating in steady state, then the energy added to inductor L during the on state and the energy dissipated from inductor L during the off state should be the same. As shown in Eq. (1.60), excess energy should not be injected or taken away from the inductor without breaking the equilibrium condition. Equation (1.61) indicates that the total inductor current change after one commutation cycle should be zero.

$$\frac{1}{2}L\Delta I_{L\,\mathrm{on}}^2 - \frac{1}{2}L\Delta I_{L\,\mathrm{off}}^2 = 0 \tag{1.60}$$

$$\Delta I_{L\,\mathrm{on}} + \Delta I_{L\,\mathrm{off}} = 0 \tag{1.61}$$

The analysis assumes ideal circuit elements. Voltage V_D is zero for the diode, and there is infinite capacitance C_{filter} on output. Voltage V_{reg} would not change over the commutation cycle for the analysis. During the on state, Eq. (1.62) gives the total change of inductor current at the end of the on state. Equation (1.63) gives the total change of inductor current at the end of the off state.

$$\Delta I_{L\,\mathrm{on}} = \frac{1}{L}\int_0^{T_{\mathrm{on}}} (V_{\mathrm{sup}} - 0)\,dt = \frac{T_{\mathrm{on}}}{L} V_{\mathrm{sup}} \tag{1.62}$$

$$\Delta I_{L\,\mathrm{off}} = \frac{1}{L}\int_{T_{\mathrm{on}}}^{T} (V_{\mathrm{reg}} - 0)\,dt = \frac{\delta T}{L} V_{\mathrm{reg}} \tag{1.63}$$

$$T_{on}V_{sup} + \delta V_{reg} = 0 \tag{1.64}$$

$$DV_{sup} = -\delta V_{reg} \tag{1.65}$$

$$\frac{V_{reg}}{V_{sup}} = -\frac{D}{\delta} \tag{1.66}$$

Equations (1.64) to (1.66) could be derived to show the ratio of output voltage V_{reg} to input voltage V_{sup}. From the angle of the view of load current, the average inductor current over the commutation cycle must be equal to the load current, as shown in Eqs. (1.67) and (1.68).

$$\overline{I_L} = I_{load} \tag{1.67}$$

$$\begin{aligned}\overline{I_L} &= \int_{T_{on}}^{T_{off}} \frac{I_{off}\,dt}{T} \\ &= \frac{I_{L\max}\delta}{2} = I_{load}\end{aligned} \tag{1.68}$$

Since $I_{L\max}$ could be derived in Eq. (1.69) and δ is a function given in Eq. (1.70),

$$I_{L\max} = \frac{V_{sup} - 0}{L}DT \tag{1.69}$$

$$\delta = -\frac{V_{sup}}{V_{reg}}D \tag{1.70}$$

Equation (1.68) could be rewritten into Eq. (1.71):

$$\overline{I_L} = \frac{I_{L\max}}{2}\delta = -\frac{V_{sup}^2 D^2 T}{2LV_{reg}} = I_{load} \tag{1.71}$$

Solving the equation for V_{reg} generates Eq. (1.72).

$$\frac{V_{reg}}{V_{sup}} = -\frac{V_{sup}D^2T}{2LI_{load}} \tag{1.72}$$

Compared with Eq. (1.59) in continuous mode of the inverting converter, the discontinuous mode output voltage V_{reg} is a much more complicated function of many parameters: input supply V_{sup}, load current I_{load}, and inductor L on duty of cycle D and cycle period T. The nonlinear relationship determines that regulation control is much more complicated for discontinuous mode than continuous mode.

1.3 Charge Pump

The charge pump is a special type of DC-DC converter. The charge pump is normally integrated on the chip where the loading circuits are located. The charge pump relies on the switching of on-chip capacitance to convert or to condition the input power supplies into the new power supplies. There is no inductive element involved in charge pump design.

The charge pump can perform many different functions that the SMPS or linear regulator is designed for. The positive charge pump, negative charge pump, and charge pump LDO regulator are a few configurations to cover charge pump functionalities.

1.3.1 Positive Charge Pump (Step-Up Converter)

The charge pump can generate an output voltage higher than the input supply on the chip. This type of charge pump is a positive charge pump.[7] Since power conversion is to step up the output voltage from the input power supply, the output current capability would drop. The positive charge pump can achieve reasonable efficiency with moderate output current load. The optimum efficiency for a positive charge pump is normally lower than the efficiency of the SMPS.

Figure 1.26 illustrates the basic circuit diagram of the positive charge pump. Capacitors and diodes are the basic building blocks.

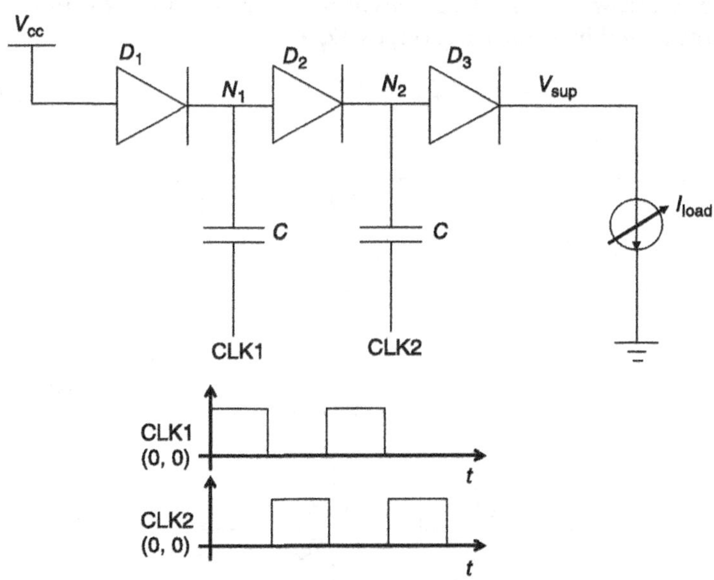

FIGURE 1.26 Positive charge pump.

Unlike the SMPS, there is no inductor involved in design. Voltage V_{cc} represents the input DC power supply. GnD is at zero potential. The output load is simplified into a variable current I_{load}. The supporting circuits for regulation are not shown. The principle of the positive charge pump is to transfer charge from the positive input supply into internal stages and then to further level-shift the potential of the transferred charge up each stage with pure capacitive coupling. Precharging and level shifting occur in two alternating phases of pump clock at each pump stage.

The operations of the positive charge pump could be explained in charge and potential point of view. Figure 1.27 illustrates pre-charging phase of positive charge pumps at stage 0. Current flows in the direction outlined by the dashed arrows. Assuming that before switching CLK1 is at V_{cc}, there is no current flow in the circuits. After CLK1 switches from V_{cc} to 0 V, the potential voltage on node N_1 would go low. The instant voltage depends on the initial voltage, the coupling ratio, and clock amplitude. Input supply V_{cc} will charge up capacitance C gradually over time through diode D_1. As the potential rises on node N_1, current will gradually decrease. The maximum voltage node N_1 could reach is V_{cc} if an ideal component is assumed where $V_D = 0$ for the diode. During the precharging phase, there is no current flow from stage 0 to stage 1 through diode D_2, since it is reverse-biased.

Figure 1.28 illustrates the level-shifting phase of stage 0 of the positive charge pump. Current flows in the direction outlined by the dashed arrows. Before CLK1 switches from low to high, the total charge stored in stage 0 is given by Eq. (1.73)

$$Q_{N_1} = V_{cc}C \tag{1.73}$$

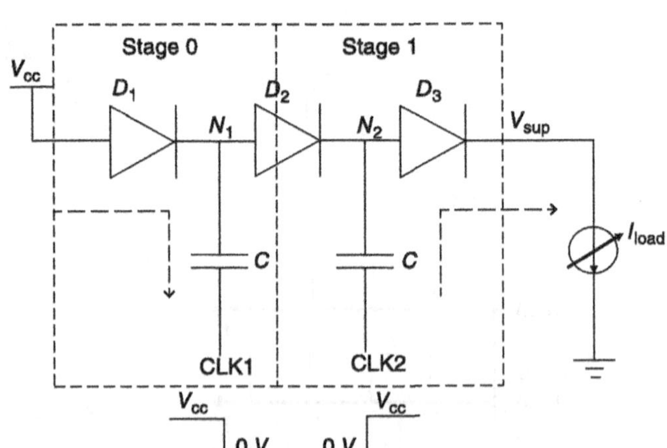

FIGURE 1.27 Positive charge pump: precharging phase of stage 0.

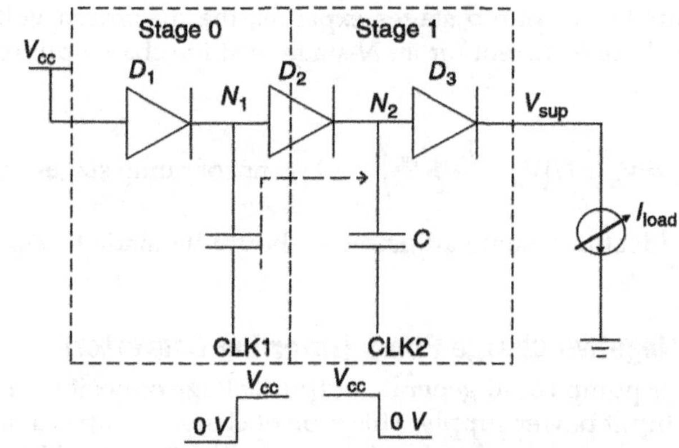

FIGURE 1.28 Positive charge pump: level-shifting phase of stage 0.

After CLK1 switches from 0 V to V_{cc}, and CLK2 switches from V_{cc} to 0, charge will be moved from node N_1 to node N_2. Diode D_1 is reverse-biased. The maximum potential that could be achieved on node N_1 is seen in Eq. (1.74); C_p is the total parasitic capacitance located on node N_1.

$$V_{N_1\,\max} = V_{cc}\left(1+\frac{C}{C+C_p}\right) \tag{1.74}$$
$$V_{N_1\,\max} \approx 2V_{cc} \qquad (\text{if } C_p << C)$$

If Eq. (1.74) is achievable, then there is no charge transfer from stage 0 to 1. Node N_2 will be minimum $2V_{cc}$ level if all components are ideal. Ideally under equilibrium conditions, the number of pump stages determines roughly the maximum voltage level that could be achieved by positive charge pump is given by Eq. (1.75)

$$V_{\max} = nV_{cc} \qquad n = \text{no. pump stages} \tag{1.75}$$

Since there is normally a load current on the output, the maximum voltage that nodes N_1 and N_2 could achieve is $V_{N_1\,\text{real}}$ in equilibrium. The excess charge given in Eq. (1.76) is used to supply the load current I_{load} within T_{clk} duration, as shown in Eq. (1.77).

$$Q_{\text{tran}} = C(2V_{cc} - V_{N_1\,\text{real}}) \tag{1.76}$$

$$I_{\text{load}} = \frac{Q_{\text{tran}}}{T_{\text{clk}}} = \frac{C(2V_{cc} - V_{N_1\,\text{real}})}{T_{\text{clk}}} \tag{1.77}$$

Likewise, if load current is known, then the maximum voltage in stage 0 is given in Eq. (1.78).

$$V_{N_1}(I_{\text{load}}) = V_{cc} + \left(V_{cc} - \frac{I_{\text{load}}T_{\text{clk}}}{C}\right) \tag{1.78}$$

As the number of pump stages expands, the maximum voltage at output with load current for an N-stage positive charge pump could be predicted as

$$V_{max} = V_{cc} + N\left(V_{cc} - \frac{I_{load}T_{clk}}{C}\right) \qquad N = \text{no. of pump stages} \qquad (1.79)$$

With nonidealities, some adjustment should be made to Eqs. (1.76) to (1.79).

1.3.2 Negative Charge Pump (Inverting Converter)

The charge pump could generate output voltage opposite in polarity from the input power supply. This type of charge pump is a negative charge pump. The amplitude of the output voltage could be either lower or higher than that of the input supplies, except the polarity is inverted from input supply. Similar to the positive charge pump, the negative charge pump can achieve reasonable power efficiency.

Figure 1.29 illustrates the basic circuit diagram of a negative charge pump. The capacitor and diodes are the basic building components. There is no inductor involved in power conversion. Voltage V_{cc} represents positive input DC supply. The GnD supply is at zero potential. The output load is simplified into a variable current load I_{load}. The supporting circuits for regulation are not shown. The operations of the negative charge pump are to discharge excess charge in each stage into ground in first clock phase, and further level shift down the potential of transferred charge with capacitive coupling in

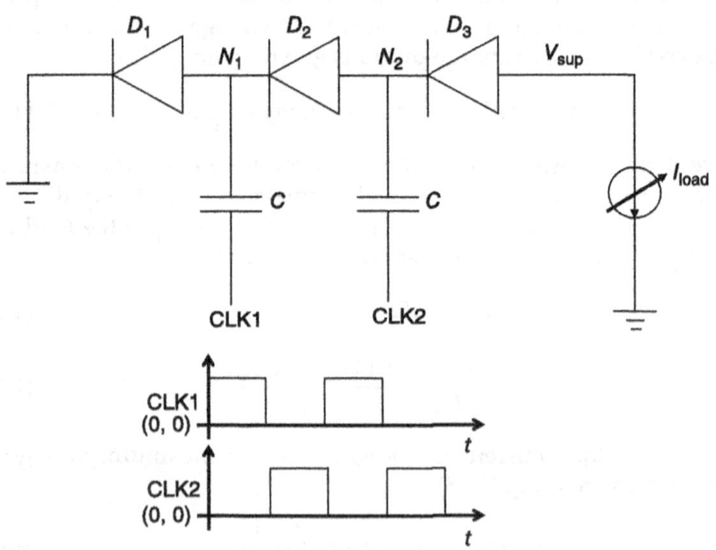

FIGURE 1.29 Negative charge pump.

the second clock phase. Discharging and level shifting occur at two alternating phases of the pump clock within a pump stage.

The principle of the negative charge pump could be explained through the charge and potential point of view. Figure 1.30 illustrates the discharging phase of the negative charge pump in stage 0. Current flows in the direction outlined by the dashed arrows. We assume that CLK1 is at 0 V, no current flows in the circuits. After CLK1 switches to V_{cc}, the potential of node N_1 goes high. Excess charge on node N_1 would be removed through diode D_1 to ground. The maximum voltage at node N_1 could reach is 0 V if we assume all ideal components where $V_D = 0$ for diodes. During the discharging phase, there is no current flow from stage 1 to stage 0 through diode D_2, since it is reverse-biased.

Figure 1.31 illustrates the level-shifting phase of stage 0 for the negative charge pump. Current flows in the direction outlined by the dashed arrows. Before CLK1 switches from high to low, the total charge stored in stage 0 is given by Eq. (1.80)

$$Q_{N_1} = (0 - V_{cc})C = -V_{cc}C \tag{1.80}$$

After CLK1 switches from V_{cc} to 0 V and CLK2 switches from 0 to V_{cc}, charge will be moved from node N_2 into node N_1. Diode D_1 is reverse-biased. The minimum potential that could be achieved on node N_1 is shown in Eq. (1.81); C_p is the total parasitic capacitance located on node N_1.

$$\begin{aligned} V_{N_1\,\text{max}} &= -V_{cc}\left(\frac{C}{C+C_p}\right) \\ V_{N_1\,\text{max}} &= -V_{cc} \qquad \text{if } C_p \ll C \end{aligned} \tag{1.81}$$

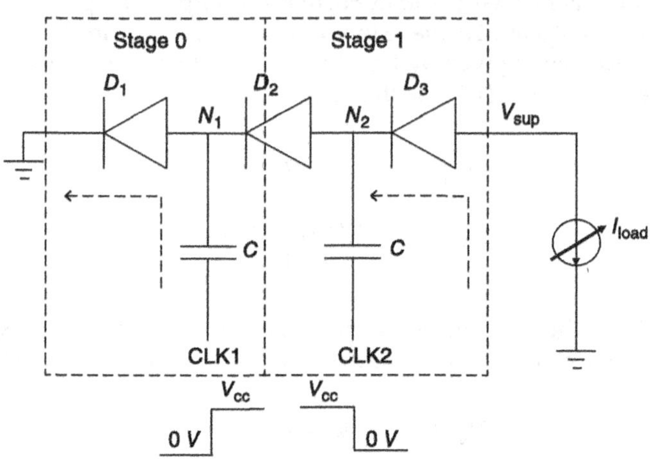

Figure 1.30 Negative pump: discharging phase of stage 0.

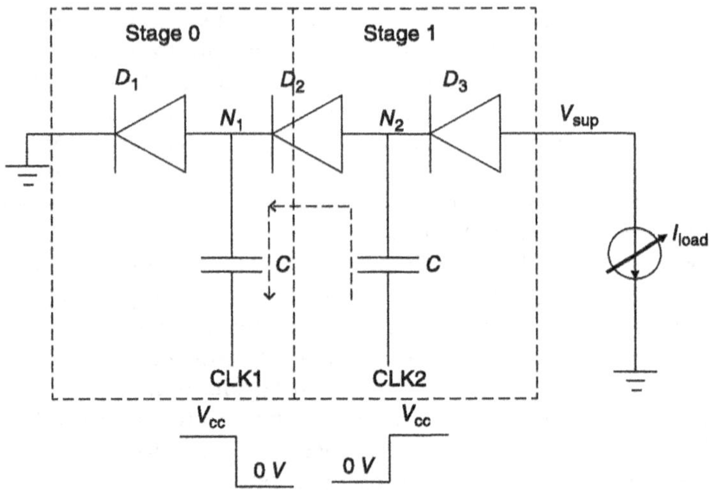

FIGURE 1.31 Negative pump: level-shifting phase of stage 0.

If Eq. (1.81) were achievable, there would be no charge transfer between stage 1 and stage 0. Node N_2 will be at maximum $-V_{cc}$ level if all components are ideal and assuming $V_D = 0$ for diodes. Ideally under equilibrium conditions, the number of pump stages determines roughly the minimum voltage level, given by Eq. (1.82), that could be achieved by the negative charge pump:

$$V_{min} = -N(V_{cc}) \qquad N = \text{no. of pump stages} \tag{1.82}$$

Since there is normally a load current on output, the potential that both N_1 and N_2 could achieve in the real case is $V_{N_1\ real}$ in steady state. The excess charge given by Eq. (1.83) is used to supply the load current defined by Eq. (1.84) with period T_{clk}.

$$Q_{tran} = C(-V_{cc} - V_{N_1\ real}) \tag{1.83}$$

$$I_{load} = \frac{Q_{tran}}{T_{clk}} = \frac{C(-V_{cc} - V_{N_1\ real})}{T_{clk}} \tag{1.84}$$

Likewise, if the load current is known, it could be easily derived that the minimum voltage could be achieved on node N_1 with Eq. (1.85).

$$V_{N_1}(I_{load}) = -V_{cc} - \frac{I_{load}T_{clk}}{C} \tag{1.85}$$

As the number of pump stages expands, the minimum output voltage at output with load current for an N-stage negative charge pump could be predicted as

$$V_{max} = N\left(-V_{cc} - \frac{I_{load}T_{clk}}{C}\right) \qquad N = \text{no. of pump stages} \qquad (1.86)$$

With nonidealities, some adjustments could be made easily to Eqs. (1.83) to (1.86).

1.3.3 Charge Pump LDO (Step-Down Converter)

With switching, the charge could be transferred from the input supply to the internal supply. Similar to the SMPS buck converter, the on-chip charge pump LDO (step-down converter) could be implemented by controlling the duty cycle of switching in operation.

Figure 1.32 illustrates the basic circuit diagram of the charge pump LDO. The switch and filter capacitance form the basis of the design. Voltage V_{cc} represents the input DC supply, and the output load is simplified into a variable current load I_{load}. The supporting circuits for regulation are not shown. The operation of the charge pump LDO is to transfer charge from V_{cc} to internal supply V_{sup} in the first clock phase. Any excess charge that could not be consumed by the load would be stored on filter capacitance C. In the second clock phase, after the switch is opened, only the filter capacitance C would supply the load.

The principle of the charge pump LDO could be explained through the charge and voltage points of view, assuming the charge LDO is operating in steady state.

Figure 1.33 illustrates the precharging phase of the charge pump LDO after the switch is closed. Current flows in the direction outlined by the dashed arrows. Ideal circuit elements are assumed, and filter C is infinite in the analysis. The current from the input supply is divided into two paths. One path is provided to meet I_{load}, and the second

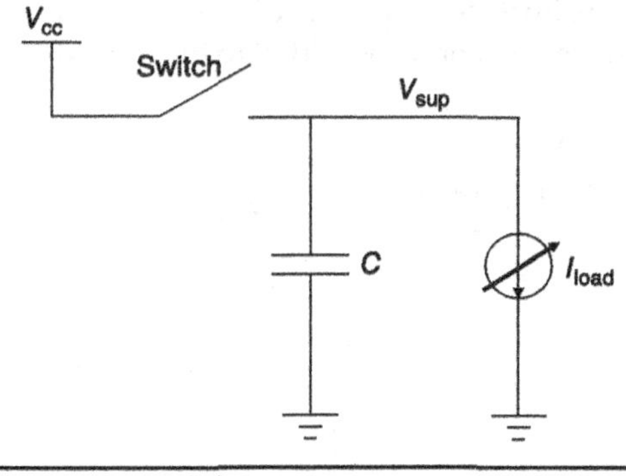

FIGURE 1.32 Charge pump LDO.

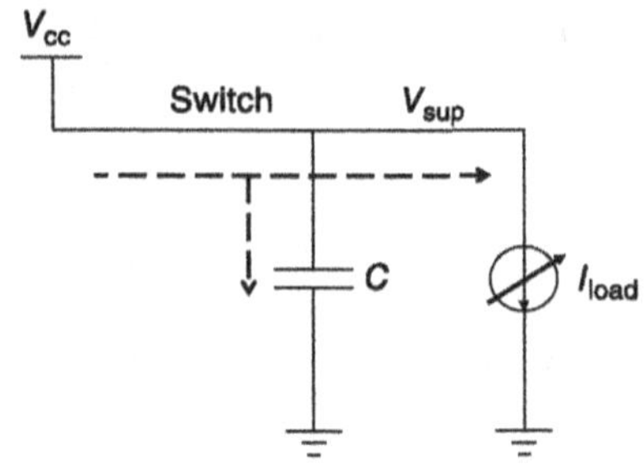

FIGURE 1.33 Charge pump LDO: precharging phase.

path is the excess supply current being stored on filter capacitance C. The stored charge is given by

$$\Delta Q_{\text{precharging}} = (I_{V_{cc}} - I_{\text{load}})T_{\text{on}} \tag{1.87}$$

Figure 1.34 illustrates the discharging phase of the charge pump LDO after the switch is opened. Current flows in the direction outlined by the dashed arrows. The only path to complete the loop of current flow is to provide current from filter capacitance C to the load. The discharged charge is given by (1.88).

$$\Delta Q_{\text{discharging}} = I_{\text{load}}T_{\text{off}} \tag{1.88}$$

If the charge pump LDO was operating in steady state, the stored charge and the charge being discharged must be equal.

$$(I_{\max} - I_{\text{load}})DT = I_{\text{load}}(1 - D)T \tag{1.89}$$

$$I_{\text{load}} = I_{\max}D \tag{1.90}$$

Equation (1.89) could be derived. Current $I_{\max}$ is the supply current available in the precharging phase. Equation (1.90) shows a similar relationship to the buck converter regulation except it is in the current domain. Since no excess charge is added to the output capacitance after

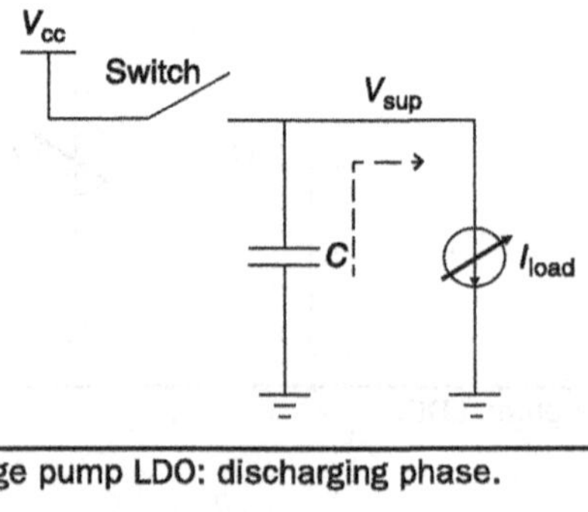

FIGURE 1.34 Charge pump LDO: discharging phase.

one commutation cycle, the output voltage would be in regulation. Negative feedback would be used to set the target voltage regulation level; then PWM should be used to maintain the charge pump LDO in regulation.

In the precharging phase, the on resistance of switch is very small since it is being fully turned on in the saturation region. The power loss on the power switch is small. In the discharging phase, there is very little leakage current through the switch since the switch is in the cutoff region. High power conversion efficiency could be achieved with charge pump LDO.

1.4 References

1. Carusone, T. C., D. Jonhs, K. Martin. *Analog Integrated Circuit Design*, 2nd ed., John Wiley & Sons, Hoboken, NJ, 2011.
2. Rincon-Mora, G. A. *Analog IC Design with Low-Dropout Regulators*, 2nd ed, McGraw-Hill Professional, New York, 2014.
3. Abraham, P., B. Keith, M. Taylor. *Switching Power Supply Design*, 3rd ed., McGraw-Hill Professional, New York, 2009.
4. Marian, G. A. *Pulse-width Modulated DC-DC Power Converters*, John Wiley & Sons, Hoboken, NJ, 2008.
5. Christophe, B. *Switch-Mode Power Supplies Spice Simulations and Practical Designs*, McGraw-Hill Professional, New York, 2008.
6. Richard, C.T. *Power converter and method for power conversion*. United States Patent Application No. US20040212357 A1. 2002, May 28.
7. Dickson, J. "On-chip high-voltage generation NMOS integrated circuits using an improved voltage multiplier technique," *IEEE J. Solid-State circuits*, vol. SC-11, no. 3, pp. 374–378, June 1976.

CHAPTER 2

Charge Pump Design Specifications and Design Metrics

To design a DC–DC converter such as a charge pump, it is important to understand the design specifications thoroughly and then be familiar with how to evaluate the performance of the charge pump with the right design metrics.

2.1 Charge Pump Design Specifications

Charge pump design specifications determine the choices of technology, circuit architectures, and regulation schemes.

2.1.1 Technology

Technology consists of the types of transistors, capacitors, and resistors that could be chosen by the charge pump design. Transistors determine the speed, transconductance, and safe operating voltage range and chip size of the charge pump. Capacitors and resistors determine chip size, the area efficiency and power efficiency of the charge pump.

2.1.1.1 Transistors

Transistors have great impact on the speed, power, and chip size of the charge pump.

For low-voltage operations, the transistor speed is strongly dependent on the transistor type and dimensions. The higher the transconductance of the transistor, the faster the speed of transition and the lesser silicon area required. The higher mobility, thinner oxide thickness, lower threshold voltage, and smaller minimum channel length L_{min} available by given technology would provide high speed and high current drivability for charge pump design.

Analog circuit design used in the charge pump requires proper V_T, g_m, R_{out}, and L_{min} to achieve high bandwidth and high gain. In most cases transistors operate under low-voltage supplies. Under special circumstances high-voltage transistors must be used to withstand high-voltage stress or biases. Those high-voltage transistors would suffer from a lower g_m with higher transistor threshold voltage, thicker transistor dielectric thickness and longer transistor channel length required to meet high voltage electrical design rules with given technology. High-voltage analog designs in the charge pump also suffer from area and accuracy issues compared with low-voltage counterparts.

For high-voltage operations, it is important to choose the right types of high-voltage transistors with proper oxide thickness, minimum allowed V_T, and minimum channel length L_{min} that withstand V_{gb}, V_{gs}, and V_{ds} in design. The threshold voltage, oxide thickness, and minimum channel length of high-voltage transistors determine the silicon area, power efficiency, transconductance, and serial impedance that could be achieved in high-voltage design.

2.1.1.2 Capacitors

The capacitor is one of the most critical circuit elements in charge pump design. The capacitance determines both the output current capability and the major contribution to the silicon area of the design.

Oxide or dielectric thickness determines the voltage range that could be safely applied between two plates of the capacitor. Once the capacitor thickness is known, the capacitance per unit area would be fixed. Output power capability and power efficiency determine the total size of capacitance needed, which is the majority total silicon area required by design.

The device's type of capacitor and physical layout design impact the overall power efficiency of the charge pump design. Power efficiency is a function of quiescent power. Since capacitive coupling is the mechanism that drives charge pump operations, parasitic R and C associated with the bottom plate and top plate of the coupling capacitor impact the power efficiency of design in different ways.

Total parasitic resistance associated with coupling capacitance is fixed by design rule constraints and actual interconnects and routings drawn in layout. The total resistance associated with coupling capacitance is the equivalent serial resistance (ESR). The higher value of ESR limits the frequency response of the capacitor. At lower frequency, the effective capacitance is unaffected by the ESR. As the transition frequency passes over the ESR Pole frequency given by Eq. (2.1), the effective capacitance would be attenuated at −20 dB/decade. The AC capability of capacitance degrades at high frequency and reduces the efficiency of the charge pump.

$$\omega_p = \frac{1}{R_{ESR}C} \tag{2.1}$$

Parasitic capacitance seen at the bottom plate of the coupling capacitance is determined by both device choice and actual physical layout. The device choice of either NMOS or PMOS capacitance would introduce a different type of parasitic capacitance on the bottom plate. For NMOS, the depletion capacitance between the channel and *P* substrate and the S/D P/N junction form the bulk of the parasitic capacitance seen on the bottom plate. If a PMOS device cap (or NWELL capacitance) was used, then NWELL to P-sub parasitic capacitance makes up the bulk of the parasitic capacitance on the bottom plate. Interconnect capacitances of clock signals contribute to the remaining portion of parasitic capacitance. Bottom plate parasitic capacitance would steal charge away from pump clock drivers and consume quiescent power through charging and discharging operations. In addition, it would slow down the slew rate of the pump clocks. As long as clock drivers are strong enough without power limitation, there is no other drawback besides the quiescent power and slew rate of the pump clock. Timing could be optimized through the upsizing of clock drivers and widths of pump clock signals.

Parasitic capacitance seen on the top plate of coupling capacitance not only would steal charge away, from going into output node but also would lower the potential voltage that could be reached on the internal node with lower coupling ratio CR, as shown in Eq. (2.2). This is one of the major killing factors for the power efficiency of the charge pump. Top plate parasitic capacitance is associated with device and interconnects, routing. Minimizing the total value of parasitic capacitance through careful and compact layout and size optimization would help limit the negative impact.

$$\mathrm{CR} = \frac{C}{C + C_{\mathrm{parasitic}}} \tag{2.2}$$

2.1.1.3 Resistors

The resistor is the passive element used frequently in the feedback network for setting the closed-loop gain of design. A better matching characteristic of passive elements over active elements is preferred. DC power consumption in the resistive feedback network could be the only DC loading to the charge pump. Reducing quiescent DC consumption requires large resistance in design. High sheet resistance is the preferred feature to reduce the total silicon area for resistor layout.

Sheet resistance and unit resistor layout size are critical in impacting the overall layout size beside capacitor. The temperature coefficient of the resistor normally would not be a concern since the feedback ratio is used to determine the final regulation target rather than the absolute resistance. The choice of resistors and unit resistor layout could impact local variation and global variation. The DNL and INL of the charge pump would depend strongly on the resistor characteristics and actual layout.

2.1.2 Loadings

Loadings of charge could be divided into capacitive loading and resistive loading. The type of loading would help to determine the regulation scheme, noise filtering technique, compensation scheme, etc.

Capacitive loading is probably one of the most commonly encountered loadings in charge pump design. The size of the total load capacitance and setting time specification determine the maximum output current, output noise amplitude, and compensation scheme.

The charge pump output current capability depends on the DC loading specification and setting time specification. The loading current specification is the maximum DC average output load current required at a given regulation level. In addition, there could be some DC current in the resistive feedback network. However, the magnitude of the feedback current is usually minimized in design to reduce the total quiescent power consumption. The settling time specification determines how much AC current must be provided in addition to the DC load. With large output load capacitance, AC current could dominate over DC current in design, and its magnitude is given by Eq. (2.3). Average I_{out} of charge pump is determined by Eq. (2.4). Maximum load current capability determines circuit design and trade-off between pump clock frequency, coupling capacitance size and pump clock amplitude to meet given design specification.

$$I_{AC} = C\frac{\Delta V}{\Delta t} \tag{2.3}$$

$$I_{out} = I_{DC} + I_{AC} \tag{2.4}$$

On the other hand, the capacitive load helps in terms of the load dump condition and noise filtering on the output supply of the charge pump. Load dump is the transient peak load current that could not be supplied by the charge pump instantaneously. To meet ΔV in the design specification, noise must be filtered out by the decoupling capacitance given in Eq. (2.5).

$$C_{decoupling} = \frac{I_{peak} t_{peak}}{\Delta V} \tag{2.5}$$

Because of the switching nature of the charge pump, output noise is a function of the current capability of the charge pump and the size of the decoupling capacitance. If current drivability I_{load} is fixed by design specification then, the output-switching noise ΔV_{noise} given by Eq. (2.6) over clock period T_{clk} could be filtered out only by total decoupling capacitance C_{load} on the output.

$$\Delta V_{noise} = \frac{Q_{tran}}{C_{load}} = \frac{I_{load} T_{clk}}{C_{load}} \tag{2.6}$$

Compensation for certain types of charge pump architectures with negative feedback control are required for stability. Normally it depends on the locations of POLEs and ZEROs in the closed loop of design. If the output node is dominant POLE with heavy load capacitance and with light load current, then the compensation should make all internal POLEs non-dominant, or to use lead compensation to improve phase margin of open-loop response. If the output node is non-dominant POLE with light load capacitance and heavy load current, then compensation need choose how to create a dominant POLE. The advantage of making output node the dominant POLE not only help stability of design, but also minimize transient noise when load dump occurs.

2.1.3 Input Supply Voltage Range

The voltage range of input supply V_{cc} has probably the most fundamental impact on the charge pump design.

$$I_{ds} = \frac{1}{2}\mu_n C_{ox} \frac{W}{L}(V_{gs} - V_T)^2 \tag{2.7}$$

$$g_m = \mu_n C_{ox} \frac{W}{L}(V_{gs} - V_T) \tag{2.8}$$

$$V_{ov} = V_{gs} - V_T = V_{cc} - V_T \tag{2.9}$$

The charge pump is designed to move a fixed amount of charge within a fixed clock period. Equation (2.7) is I_{ds} of the NMOS transistor in the saturation region, and Eq. (2.8) is the transconductance of the NMOS transistor operating in the saturation region. PMOS transistors have equivalent formulas to Eqs. (2.7) and (2.8). The lower bound of the input supply range determines the lowest overdrive [Eq. (2.9)] that the design can achieve. The higher the input supply V_{cc}, the larger the overdrive of transistor and the stronger the current controllability with same device dimensions. Higher g_m means that the charge pump can operate at higher frequency and smaller device sizes if needed.

As in the supply voltage is being scaled down with the scaling of logic process and technology to reduce chip power consumption and to increase total transistor counts per unit area, high-voltage design specification for the charge pump is unlikely to be scaled down owing to the voltage range requirement required by device and technology. For the design of pump clock drivers larger W/L ratio is required to compensate for the loss or reduction of V_{cc} in the overdrive voltage V_{ov}. Larger W/L means that the total silicon area for the charge pump needs to be increased to meet the same specification as for the previous generation of technology.

On one hand, the high end of the input supply range will translate to better performance but also larger output supply noise.

Special design technique should be taken to minimize the noise. On the other hand, the lower input supply range would be used to choose the device type for operations. Oxide thickness, threshold voltage V_T, and L_{min} are the key parameters for performance, as shown in Eqs. (2.7) to (2.9). Performance in return determines the total silicon area for the entire design.

In the extreme case, as the input supply V_{cc} is scaled too aggressively relative to the threshold of the transistor, charge pump design could become very inefficient in power and area. There is a technique of pump clock driver design with stacking capacitors introduced in Chap. 5 that would help to solve some of the issues.

2.1.4 Input Supply Current Capability

The input supply current capability could limit the pump performance, which in turn constrains the overall chip performance. The input current could be constrained by either the DC or AC need keep "current" limit given by system-level specification. The DC input supply current limit determines the maximum average power that could be delivered to the chip from the system. The AC limit is set at system level to the guard band of the power integrity for reliable operation. It is normally determined by the bandwidth of the system-level power supply generator and filtering capacitance required for stability and/or the load dump.

Two methods could be used to meet the input current limitations. The first solution is to make a trade-off between chip performance and chip power consumption. The performance of power-hungry blocks, such as the charge pump, plus the remaining circuits on the chip could be slowed down to lower either DC or AC current consumption in operations. The second solution is chosen to improve the overall power conversion efficiency of the charge pump and regulation circuits through careful architecture and circuit engineering. In many systems, charge pumps have a huge impact on both AC and DC current consumption in the chips where they are located, and it does require great attention in the design phase to meet the overall system power specifications and power integrity.

2.1.5 Output Supply Voltage Range

Unlike the SMPS boost converter in which the output voltage level could be easily adjusted by the duty of cycle with PWM, charge pump design requires very clear specification of the output supply voltage range. If the target output supply voltage range is determined for a given charge pump design, Eq. (2.10) is revised to include the threshold voltage V_T of an NMOS diode-connected device for ease of analysis. The maximum voltage for a positive charge pump is a function of the input supply voltage, threshold of the passing element, load current, clock frequency, and coupling capacitance.

$$V_{max} = (V_{cc} - V_T) + N\left(V_{cc} - V_T - \frac{I_{load}T_{clk}}{C}\right)$$

(N is number of pump stages) (2.10)

$$N = \frac{V_{out\ max} - (V_{cc} - V_T)}{V_{cc} - V_T - (I_{load}T_{clk})/C} \quad (2.11)$$

The minimum number of pump stages required by a given maximum output supply voltage range could be derived in Eq. (2.11). The larger the output supply voltage needed, the greater the number of pump stages required.

Normally pump clock signals in charge pump design should be driven by low-output impedance drivers with maximum strength to improve the total response time of clock signals. The supply for clock drivers is input power supply to maximize drivability. Equation (2.12) shows the potential stage gain per pump stage as a function of various parameters: input power supply, threshold voltage of diode-connected NMOS, load current, pump clock period, and coupling capacitance. The stage gain degrades dramatically as input power supply is being scaled down, while the threshold voltage of diode-connected NMOS does not scale down with technology, as shown in Eq. (2.13).

$$V_{gain} = V_{cc} - V_T - \frac{I_{load}T_{clk}}{C} \quad (2.12)$$

$$\frac{V_{gain}}{V_{cc}} = 1 - \frac{V_T + I_{load}T_{clk}/C}{V_{cc}} \quad (2.13)$$

Figure 2.1 illustrates the asymptotic behavior of stage gain versus input supply voltage V_{cc}. As input supply V_{cc} is scaled low enough,

V_{cc}	V_T	V_{gain}/V_{cc}
5	0.8	0.84
4	0.8	0.80
3	0.8	0.73
2	0.8	0.60
1	0.8	0.20

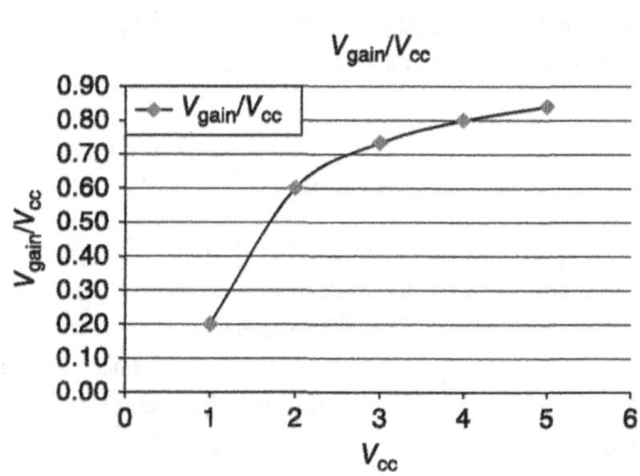

Figure 2.1 Stage gain versus input supply voltage V_{cc}.

the stage gain attenuates at a much higher rate. This illustrates the challenge of the design charge pump with low input supply.

Although the logic domain power supply is being scaled down with technology to drive thinner oxide and to reduce the total power consumption of design, the scaling of power supply is not favoring charge pump design. More pump stages are required to offset the reduction of supply voltage in order to meet an identical output voltage target range. Much larger coupling capacitance per stage is required to maintain the strength of charge pump design. Larger capacitance for the charge pump means larger silicon area with the scaling of the input power supply.

$$V_{max} = (V_{cc} - V_T) + N\left(V_{clk} - V_T - \frac{I_{load}T_{clk}}{C}\right)$$

(N is number of pump stages) (2.14)

How should one address the concern of power supply scaling impact on the charge pump? If pump clock drivers are given a new power supply V_{clk} instead of using the input power supply V_{cc} directly, then Eq. (2.10) could be revised Eq. (2.14). We assume that before the change of power supply, low V_{cc} forces the charge pump stage gain to be stuck in the region with a steeper attenuation slope. According to Fig. 2.1, if the amplitude of V_{clk} could be reasonably large enough compared to the amplitude of V_{cc}, then the new pump stage gain could be brought back into the region with a less steep slope. If the cost of generating V_{clk} was less expensive than the cost of design using lower V_{cc} directly, the new change would allow the size of the charge pump design to have lower impact from lowering the input supply V_{cc} than the original one. In Chap. 5, we discuss an approach to address this design challenge.

2.1.6 Output Supply Current Capability

The output supply current capability at target regulation is another important design specification.

$$I_{load} = \frac{C}{T_{clk}}\left[(V_{cc} - V_T)\left(1 + \frac{1}{N}\right) - \frac{V_{max}}{N}\right] \quad (2.15)$$

Equation (2.15) is derived from Eq. (2.10) to show the key parameters that could impact load current I_{load}. If the total layout size of a given design needs to be minimized, T_{clk} should be reduced proportionally to C to keep the ratio between C and T_{clk} constant. Running a pump at higher clock frequency implies that charge must be transferred at higher speed. High speed requires larger device dimension and layout size, which in turn introduce larger parasitic capacitance. Running the charge pump at higher clock frequency implies the switching power

[Eq. (2.16)] of design would increase from larger clock drivers, larger parasitic capacitance, etc. It could have a negative impact on overall power efficiency even though there is a clear gain in area efficiency.

$$P_{SW} \propto CV^2 f_{osc} \tag{2.16}$$

If the effect of the threshold voltage of passing elements could be reduced or canceled with a reasonable amount of effort in charge pump design, there would be a significant gain in output current capability in Eq. (2.15). Lowering the coupling capacitance C with V_T cancellation would achieve better power efficiency and area efficiency.

2.1.7 Output Supply Noise

Charge pump output noise is another important design specification that impacts performance and device reliability. With a linear regulator, a stable output voltage could be achieved if the system and all inputs to the system were stable. The charge pump is a switch mode power supply. Unlike a linear regulator, the nature of the switch mode itself determines that there would be noise on the charge pump output even if the systems were stable. After the switching of the pump clock edge, the system is unable to correct any error on the output until the arrival of next pump clock edge or when excess charge is completely dissipated by the output load current.

$$\Delta Q = C(\Delta V_{clk} - V_T) \tag{2.17}$$

$$\Delta V_{sup} = \frac{\Delta Q}{C_{total}} = \frac{C(\Delta V_{clk} - V_T)}{C_{total}} \tag{2.18}$$

The amount of charge delivered to the pump output within one clock edge is given by Eq. (2.17). The coupling capacitance C and voltage swing of pump V_{clk} determine the total amount of charge ΔQ being dumped to the output of the charge pump within half a clock cycle. If the transition time was short, Eq. (2.18) could approximate the maximum output supply swing after the switching of the pump clock edge. C_{total} is the sum of the load capacitance C_{load} and decoupling capacitance $C_{decoupling}$. If there was a nonideal ESR resistor associated with various capacitance, then higher output noise would be expected with the consideration of ESR resistance.

2.1.7.1 Regulation

Types of charge pump regulation schemes determine the output supply noise and power efficiency of the charge pump system.

With either on/off regulation or other types of the pulse frequency modulation (PFM) could be an optimum choice for regulation. On one hand, the output noise is large; on the other hand, it is most optimum to reduce the quiescent power consumption due to switching, to improve the overall power efficiency. If the system could tolerate large output supply noise, or if the system could afford larger decoupling capacitance C_{load} to suppress output noise, then on/off regulation would be an ideal choice for regulation due to its simplicity of design and high power efficiency.

A closed-loop system with negative feedback control on the output supply amplitude could achieve smaller output supply noise than what could be achieved by either on/off regulation or other types of PFM. In one form of implementations, the frequency of the pump clock is fixed. Negative feedback control on the pump clock amplitude is applied. While minimizing the output supply noise, the design consumes much higher quiescent switching power and has lower power efficiency.

2.1.7.2 Stability

The stability analysis of the charge pump with negative feedback is not simple. Unlike other analog circuits, AC analysis of the charge pump could not be performed in a straightforward way using circuit simulator like Spice. Theoretically, small-signal modeling could be done on the circuit elements in the entire charge pump regulation loop at first; then system-level analysis could be done to check stability. In many situations it is only possible to analyze the stability of charge pump design with transient response waveforms. From the transient response of internal nodes, it could be difficult to differentiate between switching noise and the design with a stability issue.

Output switching noise should be consistent per clock cycle under stable conditions. From the view of stability, the switching noise between two pump clock edges would be filtered out and the imaginary output should be viewed as a smooth line. The stability of closed-loop charge pump design could be analyzed with transient simulations with injection of stimulus with step function into the closed-loop system. The response of system should be checked at more relaxed resolution compared with the pump clock frequencies. Any of the analog circuit design techniques from stability analysis and compensation scheme should be able to apply to the charge pump design.

2.2 Charge Pump Design Metrics

The quality of charge pump circuit design could be evaluated against known design metrics. The evaluations could be done at design and simulation phases with proper modeling and at the characterization phase.

2.2.1 *IV* Characteristics

The *IV* curves would be the most straightforward way to evaluate the quality of charge pump design. As shown in Fig. 2.2, as a DC–DC

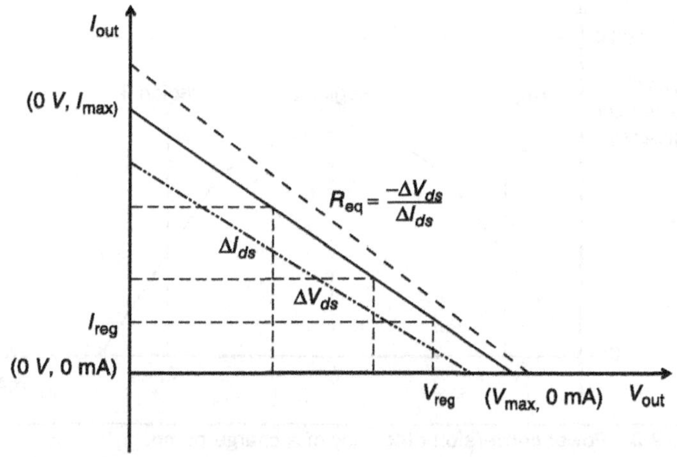

FIGURE 2.2 *IV* curves of charge pump.

power supply the output current capability at any given output voltage could be easily extracted from the *IV* curve. The slope of the *IV* curve represents the impedance of output power supply. With the variation of circuit design, physical layout, voltage, process or temperature, a family of *IV* curves could be generated to illustrate the sensitivities of given design. The maximum output voltage V_{max}, maximum output current I_{max}, impedance output supply R_{eq}, output supply current capability I_{reg} at target output regulation level V_{reg}, could be extracted from the simple graphs. Improvement of charge pump designs could be viewed intuitively and interactively through the movement of *IV* curves.

2.2.2 Charge Pump Power Efficiency

Power efficiency is another one of the most important design metrics, as shown Eq. (2.19). Here P_{out} is the total output power from the charge pump delivered to the load, and $P_{quiescent}$ is the additional power consumption in the charge pump that is not delivered to load. If P_{out} is much greater than $P_{quiescent}$, power efficiency will be high. It may not be a concern for the magnitude of $P_{quiescent}$ if P_{out} is much greater than $P_{quiescent}$ supply is large. However, if P_{out} is small or if it is comparable to $P_{quiescent}$, then power efficiency will be reduced. The charge pump designs with low load current face the greatest challenges in improving power efficiency. Particularly if the low load current weighted heavily in total operation time, the overall charge pump power efficiency could suffer if efficiency was optimized only for a high load current condition. This design metric is particularly

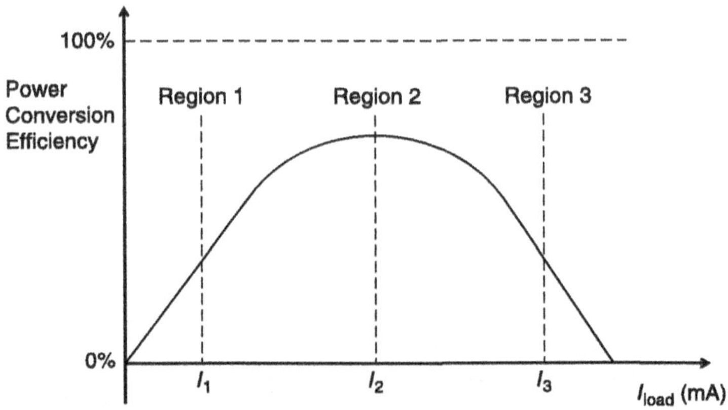

FIGURE 2.3 Power conversion efficiency of a charge pump.

critical if the chip was integrated in systems that have a tight power budget, such as handheld device, SSD, etc.

$$\text{Power efficiency} = \frac{P_{out}}{P_{out} + P_{quiescent}} \tag{2.19}$$

Figure 2.3 illustrates the power conversion efficiency[1] of a charge pump. Typically the efficiency curve peaks in the center in region 2 with output current around I_2. If the loading current increases or reduces from I_2, the power efficiency drops below the maximum power efficiency value. If I_{load} increases, internal parasitic capacitances and resistances could impact heavily the timings and voltages of power and signals. Power efficiency is reduced due to parasitic effects. If I_{load} decreases, $P_{quiescent}$ will be a significant term compared with I_{load} and will reduce power efficiency, as shown in Eq. (2.19).

Normal design would maximize the power efficiency at the peak load current since not only is it driven by timing specification, but also it would maximize power saving due to the maximum amplitude of load current. In a real operation, the probability density function of load current over the operation may suggest the case of low load current or of no load current would account for the most significant weight in total operation time. Improvement of pump operation in those corners would be as important as the maximum load current case.

At large scale, the power efficiency is set and differentiated by architectures. At small scale the power efficiency is differentiated by circuit engineering and actual physical layout.

2.2.3 Architecture

A simple comparison between the Dickson charge pump[2] and any type of V_T cancellation pump architecture could show the impact of architecture on the power efficiency.

$$I_{\text{load1}} = \frac{C}{T_{\text{clk}}}\left[(V_{cc} - V_T)\left(1 + \frac{1}{N}\right) - \frac{V_{\text{max}}}{N}\right] \tag{2.20}$$

$$I_{\text{load2}} = \frac{C}{T_{\text{clk}}}\left[V_{cc}\left(1 + \frac{1}{N}\right) - \frac{V_{\text{max}}}{N}\right] \tag{2.21}$$

$$P_{\text{out1}} = I_{\text{load1}} V_{\text{reg}} \tag{2.22}$$

$$P_{\text{out2}} = I_{\text{load2}} V_{\text{reg}} \tag{2.23}$$

Equation (2.20) shows the output current capability of Dickson charge pump with the constraints of threshold of a diode-connected transistor and other parameters. With any type of V_T cancellation charge pump architecture, with a small amount of effort, the effect of V_T should be removed, as shown in Eq. (2.21). Since both designs are targeted for identical regulation level V_{reg}, Eqs. (2.22) and (2.23) show the output power capability of two architectures at target output regulation level.

Equation (2.24) shows the power efficiency of the Dickson charge pump. The approximation is made by assuming quiescent power is negligible compared with load power. Equation (2.25) shows the efficiency for a generic V_T cancellation charge pump. The same assumption was made for simplification. Since P_{out2} is greater than P_{out1}, Eq. (2.26) illustrates the power efficiency gain from a generic charge pump with V_T cancellation over the Dickson charge pump.

$$\begin{aligned} \text{Eff}_1 &= 1 - \frac{P_{\text{quiescent}}}{P_{\text{out1}} + P_{\text{quiescent}}} \qquad (P_{\text{out1}} >> P_{\text{quiescent}}) \\ &\approx 1 - \frac{P_{\text{quiescent}}}{P_{\text{out1}}} \end{aligned} \tag{2.24}$$

$$\begin{aligned} \text{Eff}_2 &= 1 - \frac{P_{\text{quiescent}} + \Delta P}{P_{\text{out2}} + (P_{\text{quiescent}} + \Delta P)} \\ &(P_{\text{out2}} >> P_{\text{quiescent}} \text{ and } P_{\text{quiescent}} >> \Delta P) \\ &\approx 1 - \frac{P_{\text{quiescent}}}{P_{\text{out2}}} \end{aligned} \tag{2.25}$$

$$\Delta\,\text{Efficiency} = \text{Eff}_2 - \text{Eff}_1 = \frac{P_{\text{out2}} - P_{\text{out1}}}{P_{\text{out2}} P_{\text{out1}}} (P_{\text{quiescent}}) \tag{2.26}$$

2.2.4 Line Regulation

Line regulation[3] is as important a design metric for the charge pump as it is for LDO.

Compared with LDO, which could have large loop gain to suppress the input supply variation, charge pump design has much greater sensitivity with the variation of input supply at DC or low frequency. Line regulation is defined in Eq. (2.27) as the ratio between the change in output regulation level and the change in input supply voltage.

$$\text{Line regulation} = \frac{\Delta V_{reg}}{\Delta V_{cc}} \tag{2.27}$$

The contributions to line regulation of the charge pump arrive from two sources.

First, line regulation is a function of PSR of the op-amp or PSR of comparator in the regulation loop and PSR of reference voltage for regulation. If the PSR of the op-amp is not large enough, the DC output could drift at DC or low frequency. The PSR of the comparator impacts the speed of the compactor; in return the duty cycle of pulse width and average value of output voltage would be affected. If the PSR of the reference voltage for regulation suffers, the variation of reference voltage would be amplified by the closed-loop gain of pump regulation and shows up in the charge pump output voltage.

Second, line regulation is a function of the output-switching noise. As the input supply increases, the output noise amplitude increases linearly. Equation (2.28) is derived from Eq. (2.18) since V_{clk} is usually a function of input supply V_{cc}. Because of the charge is dumped to the output instantaneously at each transition edge of pump clock, only the coupling capacitance C and total capacitance at output C_{total} matter for a second source of line regulation.

$$\frac{\partial V_{sup}}{\partial V_{cc}} = \frac{C}{C_{total}} \tag{2.28}$$

The DC or low-frequency average of output switching noise from Eq. (2.28) would be added directly on top of the output supply DC average. This is the term that does not exist in LDO design. Usually the second source, Eq. (2.28), will dominate in weight in the line regulation of the charge pump. To minimize the line regulation variation, it is natural to suppress the sensitivity of pump clock V_{clk} from input supply V_{cc}.

$$\frac{\partial V_{sup}}{\partial V_{cc}} = \left(\frac{\partial V_{sup}}{\partial V_{clk}}\right)\left(\frac{\partial V_{clk}}{\partial V_{cc}}\right) \approx 0 \quad \text{if } \frac{\partial V_{clk}}{\partial V_{cc}} = 0 \tag{2.29}$$

Equation (2.29) illustrates the reduction in line regulation sensitivity if the variation of pump clock amplitude can be minimized to zero.

Due to the high-frequency nature of the clock transition, it would be better to use LDO to regulate the power supply for pump clock drivers. At the expense of power efficiency, line regulation could be significantly improved.

2.2.5 Load Regulation

Load regulation[3] is another design metric for the charge pump as the counterpart of LDO. Load regulation is defined in Eq. (2.30) as the ratio between the change in output regulation level and the change in load current at DC or the low-frequency region. It is the equivalent to the small signal output impedance of the charge pump at DC.

$$\text{Load regulation} = \frac{\Delta V_{reg}}{\Delta I_{load}} \tag{2.30}$$

Similar to the analysis of line regulation, the contribution to load regulation in the charge pump could arrive from two sources.

First, load regulation is a function of the open-loop gain of pump regulation. At DC or low frequency, as the load current varies, the open-loop gain of the system would change. Since closed-loop gain is a function of open-loop gain, with degraded open-loop gain the output voltage level would vary with the change in load current.

Second, similar to line regulation, load regulation is also a function output-switching noise. As the load current increases, the output noise amplitude would increase linearly. Equation (2.31) could be derived to show the load regulation. Here T_{clk} is duration between two successive pump clock edges, and C_{total} is the sum of the total load capacitance and the decoupling capacitance at the output. Constant k falls between 1 and the total number of pump stages N. The value is sensitive to the architecture and implementation.

$$\frac{\partial V_{reg}}{\partial I_{load}} = k\frac{T_{clk}}{C_{total}} \qquad 1 \le k \le N \tag{2.31}$$

If the load current I_{load} changed, the output voltage would change linearly. A shorter period T_{clk} can reduce the load regulation noise, and larger capacitance at the output supply can suppress load variation effectively.

2.3 References

1. Pan, F., Samaddar, T. Charge Pump Circuit Design. McGraw-Hill, New York, NY, 2006
2. Dickson, J. "On-chip high-voltage generation NMOS integrated circuits using an improved voltage multiplier technique," *IEEE J. Solid-State Circuits*, vol. SC-11, no. 3, pp. 374–378, June 1976.
3. Rincon-Mora, G. A. *Analog IC Design with Low-Dropout Regulators*, 2nd ed., McGraw-Hill Professional, New York, 2014.

CHAPTER 3

Single-Stage Charge Pump

To design a charge pump that can meet the performance specifications under given conditions (technology, supply voltage, temperature, load, etc.), there should be a systematic approach to address the design from architecture to final implementation. As a DC–DC power converter, the charge pump is considered a power supply. There are important design specifications associated with the power supply: input supply range, input current capability, output voltage range, output current capability, power conversion efficiency, etc. The charge pump needs to deliver the required output power while still being able to keep the output in regulation voltage within acceptable error tolerance limit. While converting the power from input supply to output supply, the charge pump needs to be efficient in power conversion. Power efficiency impacts the chip size, performance, battery run time, heat dissipation within a system, cost, etc. There could be different architectures that can meet the same design specification. However, power efficiency would be very different due to different architectures. Even if one uses the identical architecture, the difference in engineering practice of circuit design and physical layout could differentiate designs in performance, power conversion efficiency, chip size, etc.

To know the insights that can differentiate architectures and circuit designs, the basic modeling of charge pump design is needed to thoroughly understand the fundamental impacts in power conversion.

3.1 Basic Modeling of Charge Pump

The charge pump is a DC–DC power converter. As shown in Fig. 3.1, the charge pump or charge pump stage could be modeled as a simple Thevenin's equivalent circuit with ideal voltage source V_{in} connected in serial with Thevenin's equivalent output resistance R_{eq}. Voltage V_{in} is the maximum output voltage V_{out} the charge pump could achieve with zero output load current sourcing from V_{out} for the positive

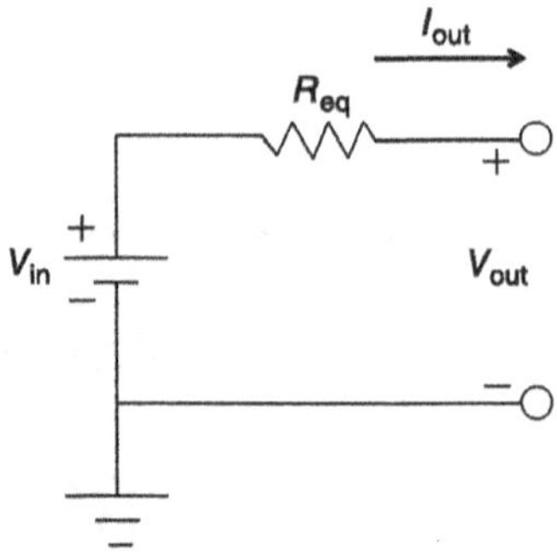

FIGURE 3.1 The venin equivalence of the charge pump.

charge pump. In terms of the negative charge pump, it would be the minimum negative voltage on the output that the charge pump could achieve with zero output load current injecting into V_{out}.

The design of the charge pump basically concerns how to maximize V_{in} and minimize R_{eq} from the given input power supply V_{cc} while being able to achieve maximum power conversion efficiency with the smallest chip size.

3.1.1 IV Curve Analysis

From Thevenin's equivalent circuits in Fig. 3.1, a few parameters could be derived for the characteristics of the charge pump or charge pump stage.

$$V_{out} = V_{in} - I_{out}R_{eq} \tag{3.1}$$

$$V_{max} = V_{in} \qquad (I_{out} = 0) \tag{3.2}$$

$$I_{max} = \frac{V_{in}}{R_{eq}} \tag{3.3}$$

$$R_{eq} = -\frac{\partial V_{out}}{\partial I_{out}} \tag{3.4}$$

From the design of point of view, few metrics are important to judge the architecture or circuit design of the charge pump or charge pump stage. Equation (3.1) shows the basic linear relationship between I_{out} and V_{out}. The equivalent output impedance of the charge pump is R_{eq}. Thevenin's equivalent voltage source V_{in} is defined in the model. By assuming zero load current, the maximum output voltage V_{max} given in Eq. (3.2) could be derived from Eq. (3.1). It is easy to see that V_{max} is equal to V_{in}. Most charge pumps are designed to supply power in either DC or AC form. Current drivability under target output voltage would determine the design capability.

Equation (3.3) defines the maximum output current I_{max} that could be achieved by shorting the output of charge pump to ground. As defined in Eq. (3.4), R_{eq} is an important parameter to judge how good a charge pump is as a power supply. The ideal power supply is expected to have zero output impedance. At a given output voltage, an infinite amount of output current could be supplied. The smaller R_{eq} is the closer the charge pump is to approach the characteristics of an ideal power supply. Both the architecture and the engineering of the design could impact R_{eq} and V_{max} of the given design. The goal of this chapter is to explore alternative charge pump architectures to maximize V_{max} and minimize R_{eq}.

The generic I_{out} versus V_{out} characteristic of the charge pump is shown in Fig. 3.2. The charge pump is simplified with linear approximation. The negative slope of the *IV* curve is the equivalent output impedance R_{eq} of the charge pump. The intercept point of the *IV* curve on the output voltage axis is the maximum output voltage V_{max} could achieve with zero load current. The intercept point of the *IV* curve on the output current axis is the maximum output current I_{max} that can be delivered to output with output node being shorted to ground.

In real applications, the charge pump is designed with closed-loop regulation. The output voltage V_{out} is being regulated in a negative feedback loop to target value V_{reg}. As shown in Fig. 3.3, as long as the output load current is within the maximum output load current capability $I_{out\ max}$ at V_{reg}, V_{out} could be regulated properly to target V_{reg}. As shown in Eq. (3.5), the output error between V_{out} and V_{reg} is inversely proportional to $1 + A_v$, and A_v is the open-loop gain of the charge pump regulation loop. If I_{out} ever exceeds $I_{out\ max}$ at V_{reg}, there will not be enough loop gain left to further correct the error. Under this condition,

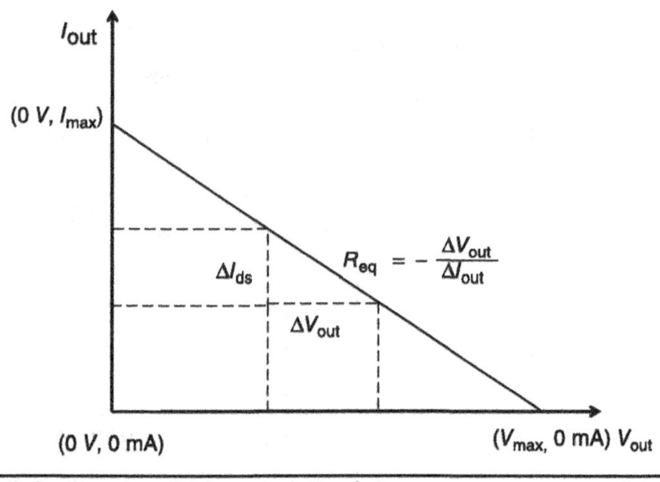

FIGURE 3.2 Open-loop *IV* characteristic of charge pump.

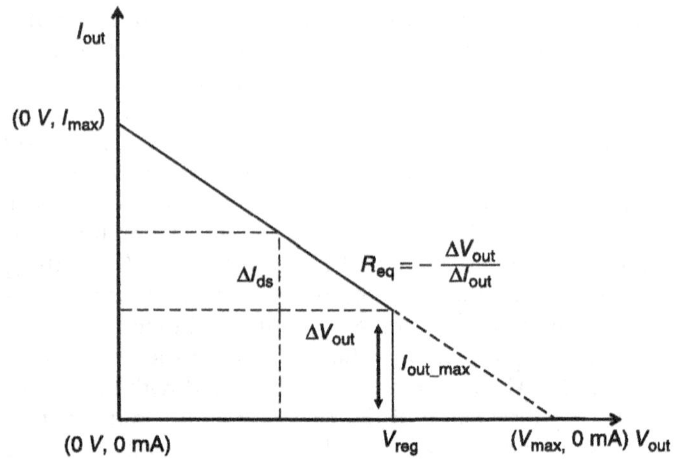

FIGURE 3.3 Closed-loop *IV* characteristic of regulated charge pump.

V_{out} could not be regulated and the output voltage could be solved by Eq. (3.1).

$$\frac{V_{out} - V_{reg}}{V_{reg}} = \frac{\Delta V}{V_{reg}} = \frac{1}{1+A_v} \tag{3.5}$$

Optimization of charge pump design is illustrated in Fig. 3.4 through the movements of *IV* curves. Two procedures could be applied to optimize the design. The first procedure is to rotate the *IV* curve in clockwise direction outlined by trend line 1 to minimize

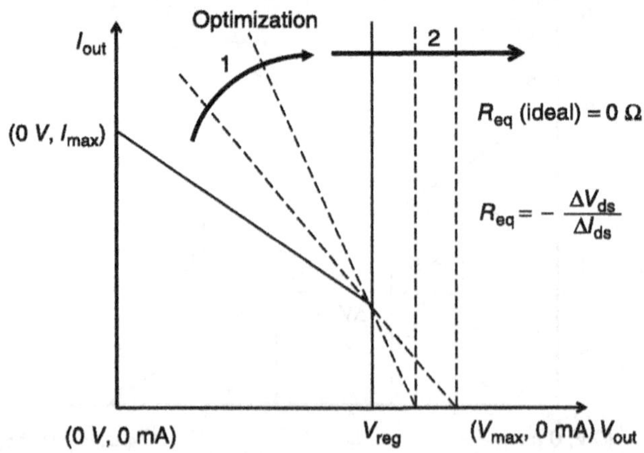

FIGURE 3.4 Optimization of charge pump design.

the R_{out} of design. With ideal power supply, R_{out} is 0Ω. The power supply could deliver an infinite amount of current at regulated voltage. In reality the design would have finite output impedance and finite power capability. The second procedure is to shift the *IV* curve in the direction highlighted by trend line 2. With identical load current, the maximum achievable regulation level is improved further.

In the following studies, the emphasis is on the architecture and circuit design for reduction of R_{out} as well as the improvement of maximum output voltage V_{max} with the area constraint.

3.1.2 Capacitive Coupling, Charge Transfer, and Diode Isolation

For an on-chip DC–DC converter, capacitive coupling is the common approach that could achieve higher potential voltage than the input power supplies given to the chip. The charge pump relies on capacitive coupling to move the potential of charge. As shown in Fig. 3.5, two capacitors are connected in series. The top plate of C_1 is connected to the input signal CLK, and the bottom plate of C_1 is connected to V_{out}. For the second capacitor, the top plate of C_2 is connected to V_{out}, and the bottom plate of C_2 is connected to ground. The center node V_{out} is initialized to V_{cc}.

The initial state of CLK is at 0 V. With a positive step function applied to CLK, how does output V_{out} respond? Figure 3.6 illustrates the step response of the system. At t', input CLK switches from 0 V to V_{cc}. Equations (3.6) and (3.7) show the transfer function of system $H(S)$ and voltage change of system ΔV_{out} with step function input. Given the initial condition that V_{out} is at V_{cc}, the condition of V_{out} at t'^{+} could be solved by Eq. (3.8). If V_{cc} were the highest input supply, V_{out} would

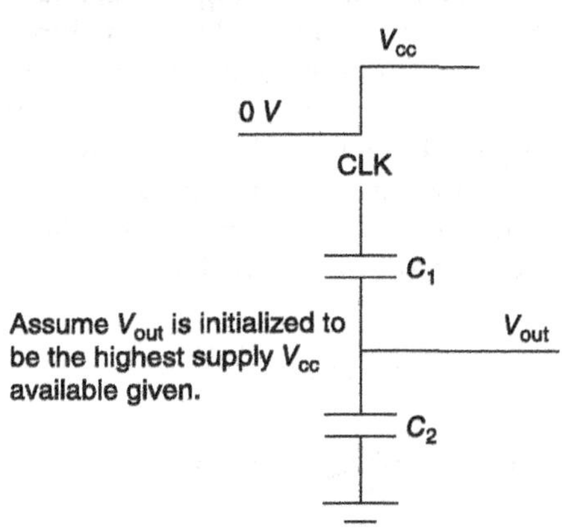

Figure 3.5 Capacitive coupling with positive clocking.

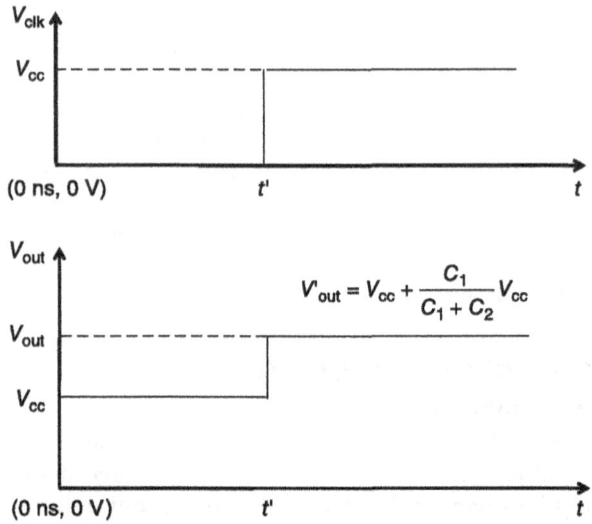

FIGURE 3.6 Waveforms of capacitive coupling with positive clocking.

be greater than V_{cc}. This is a simple illustration of how capacitive coupling could generate higher potential than input chip supplies.

$$H(S) = \frac{1/sC_2}{1/sC_1 + 1/sC_2} = \frac{C_1}{C_1 + C_2} \tag{3.6}$$

$$\Delta V_{out} = \frac{C_1}{C_1 + C_2} V_{cc} \tag{3.7}$$

$$V_{out} = V_{cc} + \Delta V_{out} = V_{cc} + V_{cc} \frac{C_1}{C_1 + C_2} \tag{3.8}$$

On the other hand, if negative voltage is needed and if the lowest input supply is at ground potential, then switching the polarity of signals could generate a negative voltage shift, as shown in Eq. (3.9). If the initial condition of the output node before switching is at 0 V, the final level of V_{out} could be solved by Eq. (3.10), which is indeed a negative potential below the lowest input supply. Figures 3.7 and 3.8 illustrate the configuration of capacitive coupling with negative clock, and waveforms associated with all nodes.

$$\Delta V_{out} = \frac{C_1}{C_1 + C_2} (-V_{cc}) \tag{3.9}$$

$$V_{out} = 0 + \Delta V_{out} = -V_{cc} \frac{C_1}{C_1 + C_2} \tag{3.10}$$

Up to this point, it is obvious that AC coupling through the capacitor could modify the potential of charge. However, if AC coupled output is connected directly to load, it could not deliver DC power to

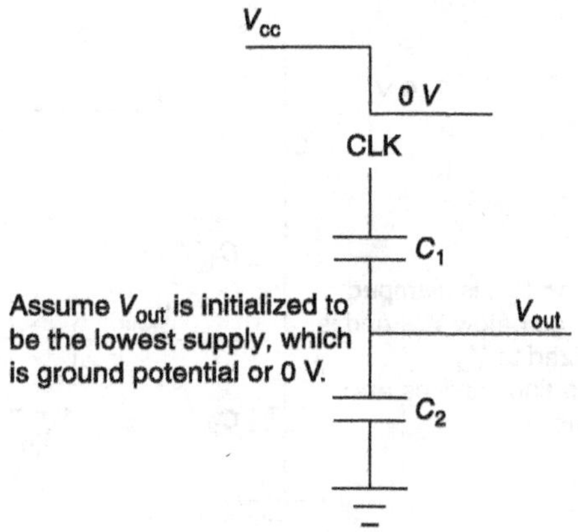

FIGURE 3.7 Capacitive coupling with negative clocking.

the load. The transient current I_{AC} that could be delivered by a capacitance is given by Eq. (3.11). There C is the size of the capacitance; V_{init} and V_{final} are the initial and final potentials, respectively, across the capacitor; and T is the duration of time for the change. Equation (3.11) could be remapped into (3.12) for a cycled event, where C is the size of capacitance, V is the potential change between cycles, and f is the switching frequency of events. The tradeoff between parameters depends upon the design specification.

$$I_{AC} = C\frac{\partial V}{\partial t} = C\frac{V_{final} - V_{init}}{T} \tag{3.11}$$

$$I_{AC} = CVf \tag{3.12}$$

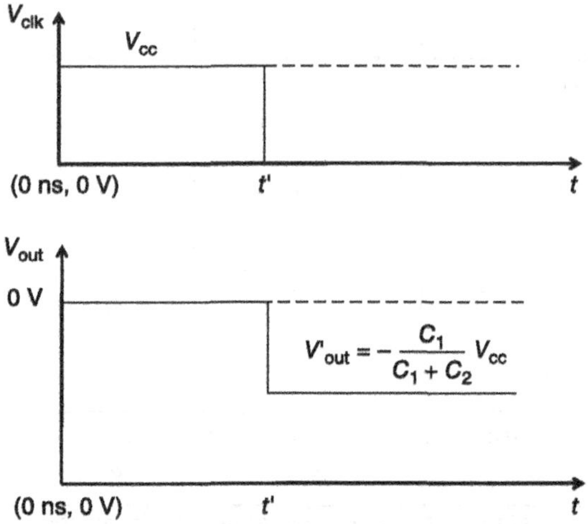

FIGURE 3.8 Waveforms of capacitive coupling with negative clocking.

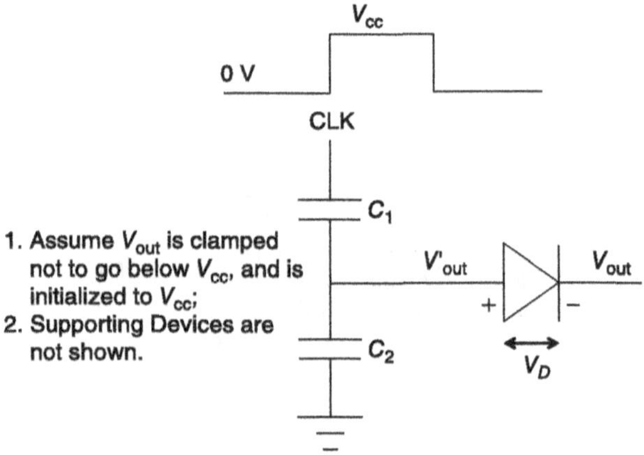

FIGURE 3.9 Charge transfer and diode isolation.

With the addition of diode, current to load is ensured to flow in one direction. Figure 3.9 shows the addition of the diode to the capacitive coupling elements, and load is isolated from the internal coupling node. V'_{out} node is clamped not to go below V_{cc} and is initialized to V_{cc}. The supporting circuits are not shown.

Figure 3.10 illustrates the rectification effect of diode, where V'_{out} is the node being AC-coupled and V_{out} is output node, which is isolated from V'_{out} by diode. The turn-on voltage of the diode is V_D. As illustrated in Fig. 3.10, in regions 1 and 2, V_{out} is one diode voltage

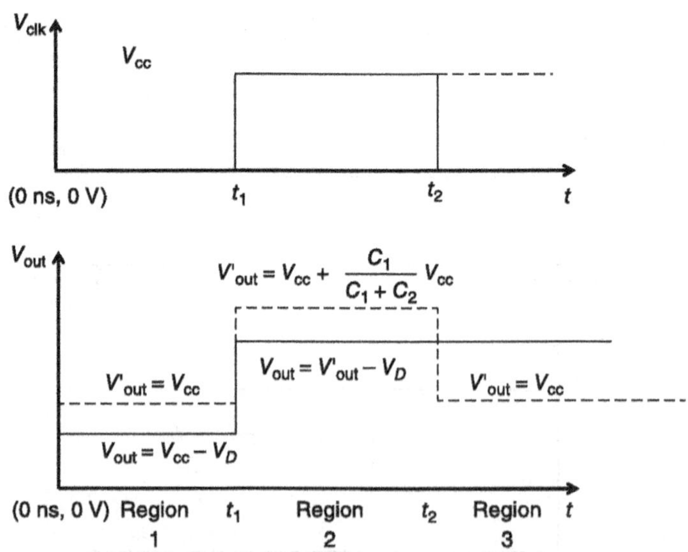

FIGURE 3.10 Waveforms of charge transfer and diode isolation.

drop V_D below V'_{out}. In region 3, as internal node V'_{out} is being coupled down by capacitive coupling, V'_{out} is clamped to V_{cc}. Because of the isolation of the diode, V_{out} will stay at higher potential than V'_{out} after it is being coupled down by V_{clk}. In region 3, the diode prevents reverse leakage of charge from V_{out} to V'_{out}.

There are a few ways to implement diodes or diode equivalents in CMOS technology. Figure 3.11 illustrates parasitic diodes that could be realized in CMOS technology. The common silicon substrate is the P substrate, and it is commonly biased to the lowest input supply, which is 0 V. Since P^+ substrate tap is tied to 0 V, it could not be part of diodes for the charge pump design. NWELL is normally formed in the P substrate. Inside the NWELL P^+ the junction could be implanted, while N^+ WELL contact is connected to NWELL. Then the P^+/NWELL diode would form a diode. Since the potential of NWELL is always greater than 0 V, the parasitic P substrate/NWELL diode is always reverse-biased.

The parasitic P^+/NWELL diode does provide very nice isolation. However, it occupies large layout area, and V_D = 0.5~0.7 V is too large if one is considering the scaling down of the power supply. The diode has a very strong relationship between I_D and V_{be}, as shown in Eq. (3.13). In Eq. (3.14), g_m of the diode has simple dependency between I_D and V_T. Equations (3.13) and (3.14) show superior current drivability, and the time constant associated with charge transfer using the diode could be very small. This small time constant is ideal for pump design. However, as shown in Fig. 3.11, available parasitic diodes in CMOS suffer from large parasitic and large silicon areas.

$$I_D = I_s e^{V_{be}/V_T} \tag{3.13}$$

$$g_m = \frac{I_D}{V_T} \tag{3.14}$$

MOSFETs could be configured as diodes with the drain and gate of MOSFET transistor connected together as illustrated in Figure 3.12.

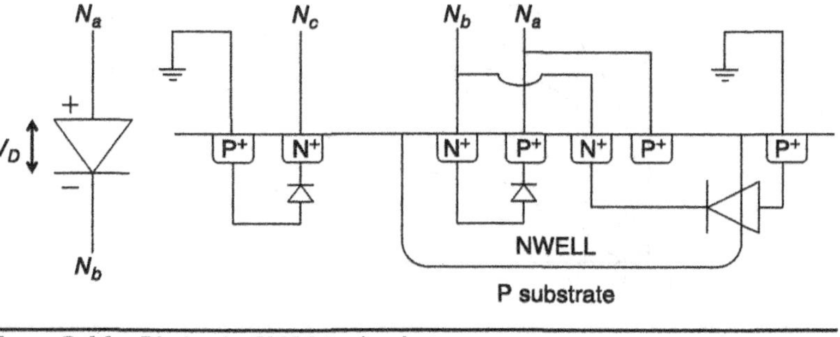

Figure 3.11 Diodes in CMOS technology.

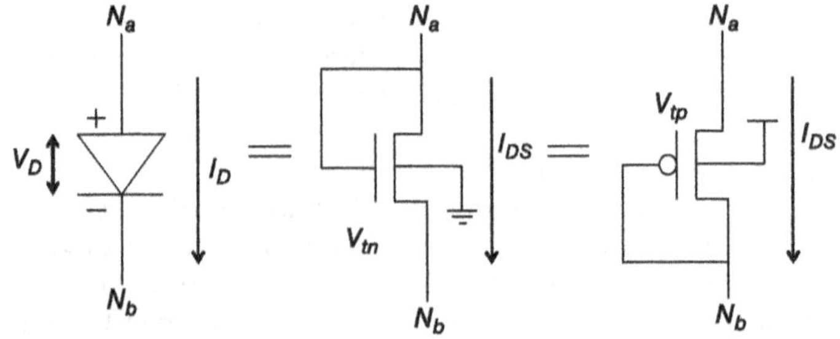

FIGURE 3.12 Diode equivalences in CMOS technology.

Since NMOS and PMOS are complementary to each other, the following analysis would focus on NMOS-related design only for simplicity.

$$I_{ds} = \frac{1}{2}\mu_n C_{ox} \frac{W}{L}(V_{gs} - V_T)^2 (1 + \lambda V_{ds}) \tag{3.15}$$

$$g_m = \mu_n C_{ox} \frac{W}{L}(V_{gs} - V_T)(1 + \lambda V_{ds}) \tag{3.16}$$

For diode-connected MMOS operations in the saturation region, I_{ds} of NMOS operates in the saturation region as described by Eq. (3.15) with respect to V_{gs} and V_{ds}, g_m of the transistor is described by Eq. (3.16). Compared with parasitic diode characteristics in Eqs. (3.13) and (3.14), diode-connected NMOS has both I_{ds} and g_m as weak functions of V_{gs}. Also I_{ds} is proportional to the second order of V_{gs}, and g_m is proportional to the first order of V_{gs}. To be able to command the same amount of current control, a larger device dimension is expected for NMOS. Figure 3.13 shows I_{ds} versus V_{ds} for different V_{gs} values. The dotted lines are generic IV characteristics of NMOS. Since gate and drain are

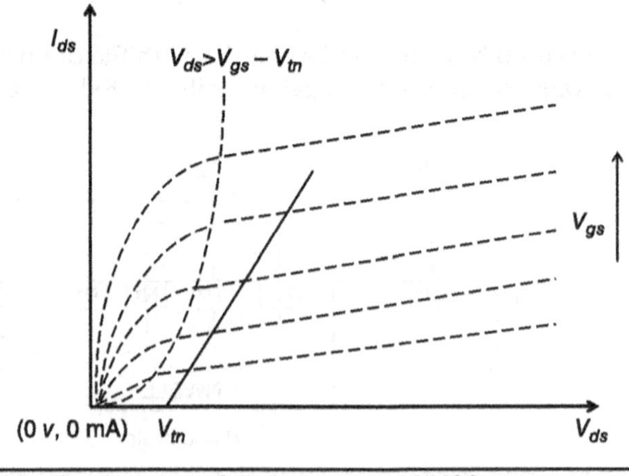

FIGURE 3.13 The IV curves of diode-connected NMOS.

tied together as the input terminal, and the source is the output terminal, the *IV* curve of diode-connected NMOS operates on the solid line, which intercepts the V_{ds} axis at $V_{ds} = V_{tn}$.

Using diode-connected NMOS, the source node of NMOS is biased at higher potential than that of the substrate. Threshold voltage V_T is a function of the source voltage as given in Eq. (3.17). With body bias, as the source voltage increases, the threshold voltage of NMOS increases.

$$V_T = V_{to} + \gamma(\sqrt{V_{sb} + 2\theta} - \sqrt{2\theta}) \tag{3.17}$$

To meet the same current drivability as the diode counterpart, diode-connected NMOS can only increase the W/L ratio to increase the g_m of the transistor. As an outcome, it requires large chip size and in addition introduces large parasitic capacitance associated with design and layout. The impact on the charge pump is two-fold. One impact is the reduction of the coupling ratio (CR) with added parasitic capacitance. The second impact is the increment of the time constant of charge transferring with added parasitic resistance.

3.1.3 Charge Pump Design with V_T Cancellation

As shown in early analysis, the implementation of the isolation transistor has a significant impact on circuit performance of the charge pump. To accomplish the same job, a larger g_m, a smaller silicon area, and less parasitic capacitance are preferred design metrics in choosing the implementation of the isolation device.

If one is choosing a parasitic diode in CMOS technology as an isolation transistor, the parasitic diode does have larger g_m compared to other types of transistors. The voltage V_D of the diode would not be a concern if the input supply were high. Following the scaling trend of input supply voltage with technology, the V_D of the diode eventually would be a concern. As shown in Fig. 3.11, the impact area of using a parasitic diode is large. Each diode needs to be located in a separate NWELL, and design rules could make the total layout size prohibitive. Associated with large NWELL, the total parasitic capacitance between NWELL/P substrate parasitic capacitance would be large, and longer interconnect would be needed. Large parasitic capacitance has a negative impact on many design metrics.

If one is working with a CMOS transistor, the g_m of the transistor is not as good as that of BJT or diode. Shorter channel length and larger W/L ratio could be used to increase g_m and the current drivability. Equivalent to V_D of the diode, the diode-connected NMOS V_D depends on the threshold voltage V_T of the transistor and the overdrive voltage V_{ov} given in Eq. (3.18). Voltage V_{ov} is a function of the transistor current defined in Eq. (3.19). If a smaller W/L ratio were used, a large V_{ov} would be developed. As a consequence, less charge would be transferred across the isolation device within a given time window. If a larger W/L ratio were used, a smaller V_{ov} would be developed.

More charge could be transferred between stages within a given time window. Another benefit of increasing the W/L ratio is the reduction of the equivalent R_{on} of the passing element defined in Eq. (3.20). Lowering R_{on} reduces the transferring time constant of charge stages and improves the transition speed.

$$V_g - V_s = V_T + V_{ov} \tag{3.18}$$

$$V_{ov} = \sqrt{\frac{2I_{ds}}{\mu_n C_{ox} W/L}} \tag{3.19}$$

$$R_{on} = \frac{1}{\lambda I_{ds}} = \frac{1}{\lambda[\frac{1}{2}\mu_n C_{ox} W/L(V_{gs} - V_T)^2]} \tag{3.20}$$

If one is using a CMOS transistor as the diode-connected device, the lower limit of voltage drop between the drain and the source in Eq. (3.18) will be $V_{gs} = V_T$. This may be okay if the input supply voltage was large or the output voltage was not too high. With the scaling of input supply voltages with technology, the impact of V_T is obvious. On the other hand, if the output voltage was large, Eq. (3.17) predicts the larger V_T of the diode-connected transistor due to the increased body bias V_{sb}. If charge could not be fully transferred per clock cycle and a portion of charge was trapped due to V_T and V_{ov}, the power efficiency of the charge pump would be lower than desired. This is the motivation for trying to cancel V_T and to overcome V_{ov} in design.

3.1.4 IV Curve of V_T Cancellation Scheme

There are many different kinds of V_T cancellation schemes. Some are easy to follow, and others are difficult to implement. All follow the similar principle illustrated in Fig. 3.14. The diode-connected NMOS IV

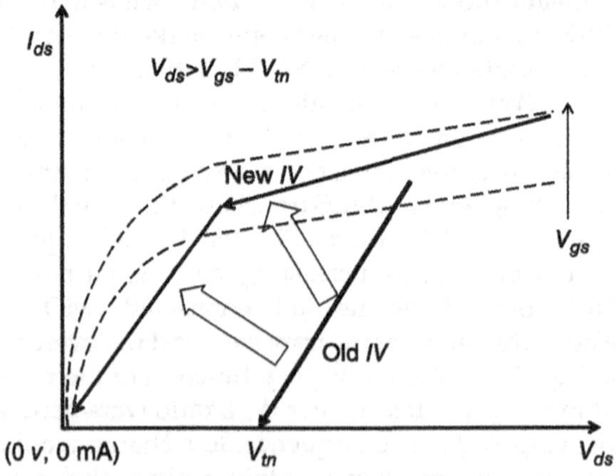

Figure 3.14 The *IV* curve of V_T cancellation scheme.

curve is labeled Old IV. It intercepts the V_{ds} axis at point $V_{ds} = V_{gs} = V_{tn}$. To fully transfer charge through NMOS without reservation, V_{ds} has to reach near 0 V at the end of the charge transfer period, at which point I_{ds} is near zero. If one is referring to the I_{ds} versus V_{ds} curves for different V_{gs} of NMOS, in order for both V_{ds} and I_{ds} to reach 0, the *IV* curve of NMOS for new design must behave similarly to the curve labeled New as *IV*. With the V_T cancellation scheme, the gate voltage of NMOS must be biased high enough to ensure the transistor is fully conducting regardless of the regions of operation of transistor. This requirement is shown in Eq. (3.21). The transistor could traverse through saturation and the linear regions in operation, or only the linear region in an extreme case.

$$V_g(t) - V_s(t) > V_T + V_{ov} \tag{3.21}$$

In actual practice, quite a few architectures could be used to illustrate the concept of V_T cancellation. From the simple approach of bootstrapping in all NMOS output drivers, to the sophisticated four-phase charge pump, the effect of the pass element's V_T is canceled.

In Sec. 3.2 the basic cross-coupled pair voltage doubler configured as a single-stage pump is discussed at the beginning to illustrate the concept of V_T cancellation. In later sections, three new V_T cancellation single-stage pump architectures evolved from the basic cross-coupled pair voltage doubler are introduced to show the simplicity and effectiveness of new designs.

3.1.5 Charge Transfer

Equation (3.11) is repeated here as Eq. (3.22) for the ease of discussion. The AC current through coupling capacitance is used to meet the load requirement within a given time period. Since switching of capacitors requires resetting of internal nodes, there could be two choices to design the charge pump with the consideration of output noise.

$$I_{AC} = C\frac{\partial V}{\partial t} = C\frac{V_{final} - V_{init}}{T} \tag{3.22}$$

If the charge pump delivers charge to the output only in the first half clock cycle, and in the second half clock cycle the internal nodes of the pump are in the reset phase, there must be enough decoupling capacitance at the output to supply the charge in the reset phase. The decoupling capacitance C must absorb any excess energy from the charge pump within the first half cycle in order to provide power solely by itself in the second half cycle.

$$\int_0^{2T} I_{load}\, dt = \int_0^{T} I_{pump}\, dt \tag{3.23}$$

$$Q_{cap} = C\Delta V_{out} = \int_0^{T} (I_{pump} - I_{load})\, dt = \int_T^{2T} I_{load}\, dt \tag{3.24}$$

Equation (3.23) shows that the power delivered to the load within an entire clock cycle is coming from only the first half clock cycle from the charge pump. T is a half cycle period, and $2T$ is a full clock cycle period. Equation (3.24) shows that any excess charge from the pump in the first half clock cycle is absorbed by the decoupling capacitance C on the output, and this absorbed charge is delivered to the load in the second half clock cycle. While the charge pump is in regulation, steady-state operations determine per clock cycle the total power derived from the pump, and the total power consumed by the load should be equal. Figure 3.15 illustrates the clock phase and output current supplied by the charge pump. If $I_{\text{load ave}}$ represents the average load current, within the first half of clock cycle, the power consumed by the load is represented by the total area C. The excess power from the charge pump represented by area A is absorbed by decoupling capacitance C in the first half clock cycle. In the second half clock cycle, the total power consumption by the load is represented by total area B. Under equilibrium, the total area A must be equal to total area B. Because of the nature of single-phase charge transfer, the output of the charge pump would experience large voltage noise in operation. To minimize the voltage noise for design specification, there should be enough decoupling capacitance assigned to filter the output noise of the charge pump.

An alternative design approach could be used to minimize the charge pump output noise. As shown in Eq. (3.25), the charge pump can deliver equal amounts of charge within each half clock cycle. The corresponding clock waveforms and charge pump output current waveform are illustrated in Fig. 3.16. With this approach, single charge pump is split into two equal branches, and each supplies current to the output

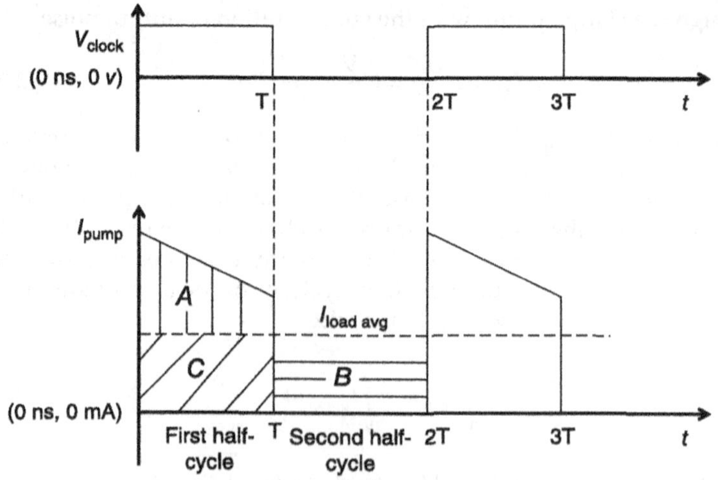

FIGURE 3.15 Charge pump with single-phase charge transfer.

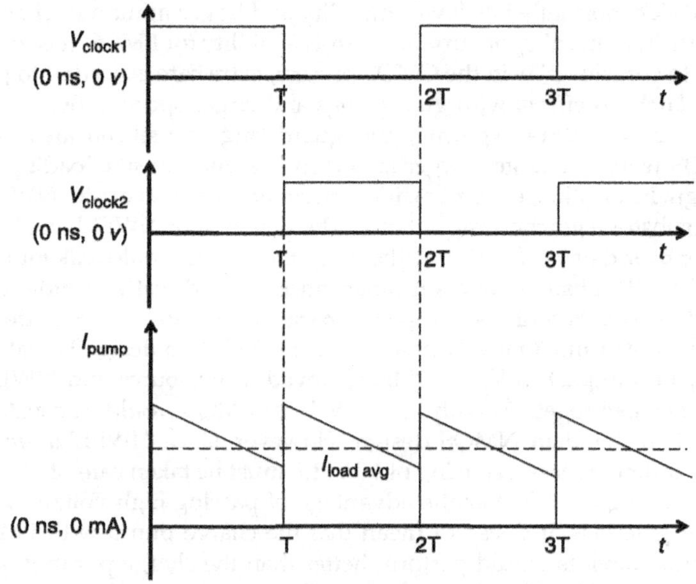

Figure 3.16 Charge pump with double-phase charge transfer.

load within alternating half clock cycles. By default, with the identical size of decoupling capacitance as in the single-phase charge transfer case, charge pump output noise would be halved automatically.

$$Q = \int_0^{2T} I_{\text{load}}\, dt = 2\int_0^{T} I_{\text{pump}}\, dt \tag{3.25}$$

To further minimize the charge pump output noise, this practice could be easily expanded into multiphase clocking schemes with multiple smaller charge pump branches to cut down output noise further. The extra cost would be the complicated clock phase control and larger chip size due to extra routings and spacing requirements.

3.1.6 P-Channel versus N-Channel Isolation Device

In the charge pump design, one can choose to use either PMOS or NMOS transistors as isolation transistors in the charge pump stage design.

NMOS normally has higher mobility and requires shorter minimum channel length. It has better g_m or current control capability. It is simple and safe to prevent latch-up related issues or concerns in operation. The drawback of NMOS is that the threshold voltage V_T is a function of V_{sb}. As V_{sb} increases, V_T of NMOS would be increased roughly by the square root of V_{sb}, and there is no magic to reduce the trend without the introduction of a triple well process.

PMOS normally has lower mobility and larger minimum channel length. In terms of g_m or current control capability for PMOS, it is weak. Due to parasitic BJTs in the CMOS process, extra care is needed to prevent latch-up effects with guard rings and larger spacing. Because of the size of NWELL, spacing, and guard ring, the silicon area with PMOS would be large. Larger area incurs larger parasitic loadings in design. In addition, extra parasitic capacitance is added due to NWELL to P substrate junction capacitance. The potential of NWELL could be either biased or left floating. Either way, the charge would leak into the NWELL/P substrate junction capacitance either directly or indirectly. All these effects could be lumped as parasitic load to internal nodes of the charge pump. One advantage of using PMOS in design is that the body bias impact to V_T could be removed if the source and NWELL could be tied together. With $V_{SB} = 0$ V, V_T of PMOS would be constant. This is better than NMOS design. However, since NWELLs would have different biases, spacing of NWELL must be taken care of.

Although PMOS has the advantage of passing high voltage with lower gate bias, it does not mean that the charge pump with PMOS isolation devices would perform better than the charge pump implemented with NMOS isolation devices. The deign is strongly technology dependent and engineering dependent. In the remaining chapters, only the charge pump with NMOS isolation devices will be used for illustration purposes. The dual of using the charge pump with PMOS isolation devices could be easily designed and analyzed since the fundamentals of charge pump design do not change.

3.2 Basic Cross-Coupled Pair Voltage Doubler

Instead of analyzing the architecture of the Dickson charge pump[1], the basic cross-coupled pair voltage doubler is used to illustrate the concept of charge pump design. The V_T cancellation schemes have gained significant popularity for higher power conversion efficiency and area efficiency in chip designs. It would be more practical to introduce circuit architectures that designers could apply easily to their designs without much struggle.

This section first introduces the basic cross-coupled pair voltage doubler. Then it is followed by three sections of new pump architectures that encompass essential design concepts and techniques of modern charge pump designs. These structures could be easily applied or extended to practical design work to meet different design specifications.

3.2.1 Introduction

Figure 3.17 illustrates the structure of the basic cross-coupled pair voltage doubler. It consists of two NMOS transistors N_1 and N_2 and two capacitors C_1 and C_2. Node n_a is connected between three nodes:

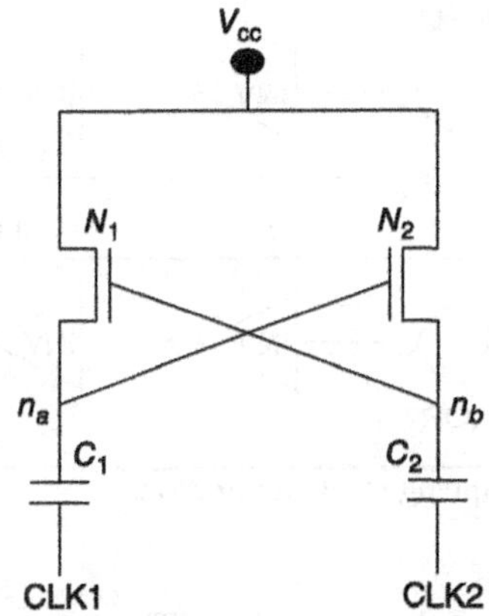

Figure 3.17 Basic cross-coupled pair voltage doubler.

the source of N_1, top plate of C_1, and gate of N_2. Node n_b is connected between three nodes: the source of N_2, top plate of C_2, and gate of N_1. Since both source and gate nodes of N_1 and N_2 are cross-coupled, it got its name from the special connection. The input power supply to the circuits is V_{cc}, which is assumed to be the highest input supply available. CLK1 and CLK2 are pump clocks that drive the bottom plates of C_1 and C_2. This cross-coupled pair turns out to be one of the simplest forms of V_T cancellation circuit that could be used in charge pump design.

3.2.2 Nonoverlapping Clock Phases

The operations of the basic cross-coupled pair voltage doubler rely upon positive feedback to perform precharging and V_T cancellation to achieve doubling of the input supply voltage V_{cc} on internal node n_a or n_b. Figure 3.18 illustrates the nonoverlapping clock scheme required by this design. CLK1 and CLK2 cannot have overlapping

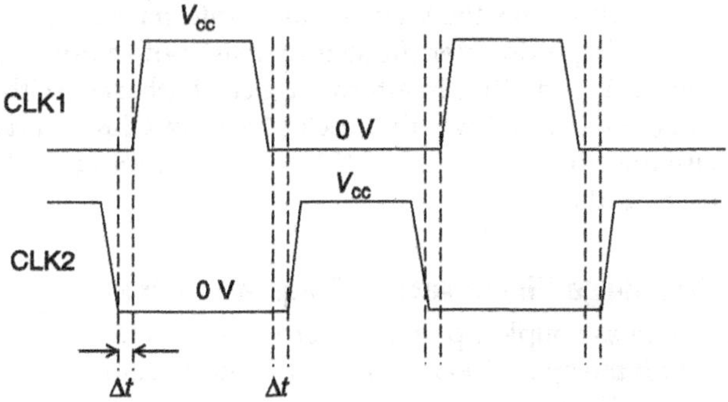

Figure 3.18 Nonoverlapping clock scheme.

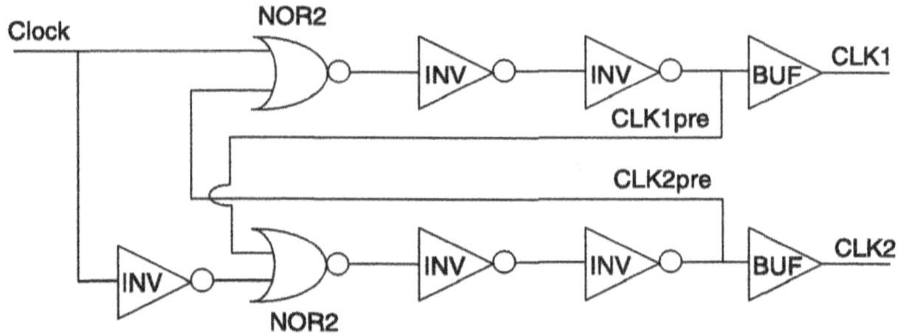

Figure 3.19 Nonoverlapping clock generation.

of high clock phases at the same time. The nonoverlapping clocking scheme prevents the boosted charge on internal nodes from leaking back into the input supply of the voltage doubler. The minimum separation between high clock phases of CLK1 and CLK2 depends on the slew rate of the clocks and V_T of N_1 and N_2. Clocks could be generated from a nonoverlapping clock generator and then buffered to drive the coupling capacitors C_1 and C_2. Because of process, voltage and temperature (PVT) variations, the design has to guarantee the minimum Δt separation margin in the worst PVT corners. For typical or other PVT corners, there could be too much timing margin left used.

Figure 3.19 shows an implementation to generate nonoverlapping clocks CLK1 and CLK2 through the system clock. With the feedback from CLK1pre and CLK2pre into NOR2 gates, nonoverlapping high phases are guaranteed. Under ideal conditions the nonoverlapping delay is the sum of gate-to-gate delays through one NOR2 and two INV. Since pump clocks normally drive a large capacitive load, buffering is required. In this example, BUFs are added at the output of CLK1pre and CLK2pre to drive CLK1 and CLK2. The added buffers could introduce additional skews, mismatches, etc. Final buffered clocks CLK1 and CLK2 could have less timing margin in nonoverlapping delay.

Figure 3.20 illustrates the waveforms of internal nodes in a cross-coupled pair voltage doubler. In steady state, both nodes n_a and n_b switch between V_{cc} and $2V_{cc}$ in alternating clock phases. Without any current loads, node n_a follows the clock phases of CLK1, and so does node n_b with the clock phases of CLK2. Nonoverlapping clock phases are held for n_a and n_b.

3.2.3 Precharge Phase versus Coupling Phase

How does a cross-coupled pair doubler work? There are two phases associated with the operations of each node: precharging and coupling.

Figure 3.21 illustrates the precharge phase of node n_a. First CLK1 switches from V_{cc} to 0 V; then it is followed by switching of CLK2

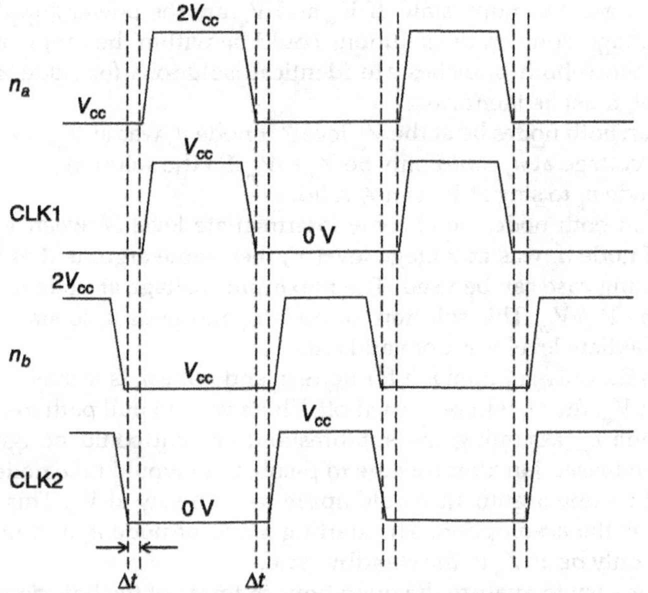

Figure 3.20 Waveforms of internal nodes in cross-coupled pair voltage doubler.

from 0 V to V_{cc}. How do nodes n_a and n_b respond in the precharge phase of node n_a? If switching happens following the standby state, could the cross-coupled pair voltage doubler be guaranteed to operate properly? Let us start by checking the possible states that nodes n_a and n_b could be

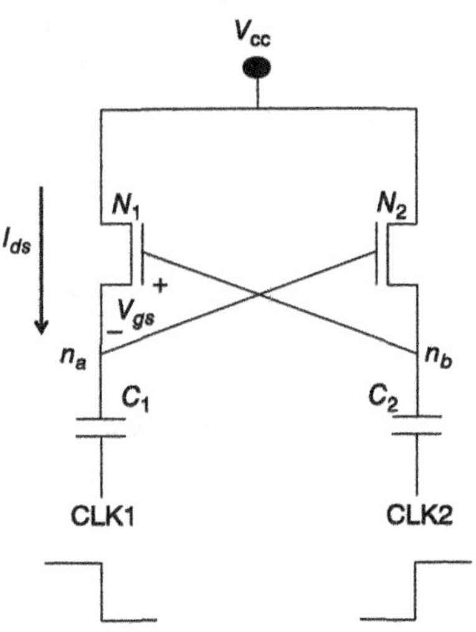

Figure 3.21 Precharge phase of node n_a.

in during the standby state. If V_{cc} and V_{ss} are the power supplies to the voltage doubler, the solutions could be within the range of supplies. Since both branches are identical, solutions for node n_a and node n_b must be identical.

Can both nodes be at the V_{cc} level? If node n_b was at V_{cc}, the maximum voltage at n_a could only be $V_{cc} - V_{tn}$. So the solution for node n_a and node n_b to stay at V_{cc} is not valid.

Can both nodes be at some intermediate level between V_{cc} and V_{ss}? If node n_b was at voltage level V, then same argument as in the preceding case can be used. The maximum voltage at node n_a could only be $V - V_{tn}$. This solution for node n_a and node n_b to stay at any intermediate level V is not valid either.

So the only solution left for node n_a and node n_b is to stay at V_{ss}. If n_b is at V_{ss}, the N_1 will be turned off. There was no pull path to charge up node n_a, assuming the subthreshold current could be ignored, reverse-biased junction leakage to p-substrate would take node n_a to V_{ss}. The same argument would apply to n_b to stay at V_{ss}. This is the same as the assumption. The starting states of node n_a and node n_b could only be at V_{ss} in the standby state.

It is easy to analyze the operations in terms of the half circuit. As illustrated in Fig. 3.22, each half circuit has one NMOS and one capacitor. The gate of NMOS is at V_{ss} in the standby state. Total parasitic capacitance terminated to V_{ss} at node n_a and node n_b is lumped into capacitance $C_{parasitic1}$ and $C_{parasitic2}$.

$$C_{parasitic1} = C_{db1} + C_{gd1} + C_{gate2} \tag{3.26}$$

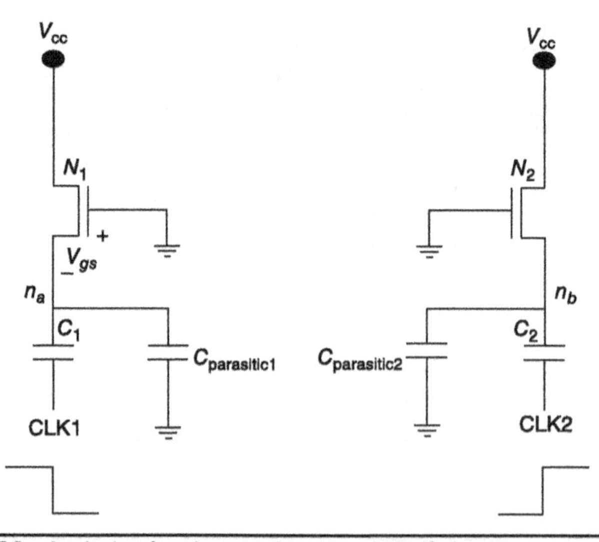

Figure 3.22 Analysis of node n_a and node n_b in half circuits.

$$C_{\text{parasitic2}} = C_{\text{db2}} + C_{\text{gd2}} + C_{\text{gate1}} \tag{3.27}$$

$$\text{CR} = \text{CR}_1 = \text{CR}_2 = \frac{C_1}{C_1 + C_{\text{parasitic1}}} = \frac{C_2}{C_2 + C_{\text{parasitic2}}} \tag{3.28}$$

Equations (3.26) and (3.27) summarize the components of parasitic capacitance in $C_{\text{parasitic1}}$ and $C_{\text{parasitic2}}$. With CLK1 switches from V_{cc} to 0 V initially, node n_a would be coupled below 0 V. Since the junction diode of N_1 connected to n_a would be forward-biased if node n_a went below $-V_D$ (−0.6 V), node n_a would be coupled to the higher value of $-V_{\text{clk}}\text{CR}$ and $-V_D$. After the nonoverlapping delay, CLK2 switches from V_{ss} to V_{cc}. Node n_b will be coupled up from V_{ss} to $V_{\text{clk}}\text{CR}$.

$$V_{\text{clk}}\text{CR} > V_{tn} \qquad (V_{sb} = V_{\text{cc}}) \tag{3.29}$$

What is the minimum requirement for a cross-coupled pair voltage doubler to function? In order for positive feedback to work, node n_a must be pulled up above 0 V by NMOS N_1 after node n_b is coupled high. Equation (3.29) gives the minimum requirement for positive feedback to start. Node n_b must be coupled higher than the threshold voltage V_{tn} of N_1 to turn on the transistor. Charge must be transferred from V_{cc} to node n_a and must pull it above V_{ss} in the precharge phase of node n_a. With positive feedback, the internal voltages of node n_a and node n_b move higher and higher in the subsequent precharge phase than in the previous cycle. In steady state, node n_a and node n_b would be at V_{cc} at the end of each precharge phase. To ensure this condition holds, Eq. (3.29) shows $V_{\text{clk}}\text{CR}$ must be greater than V_{tn} of N_1 or N_2 with source biases at V_{cc}.

$$V_{\text{clk}}CR < V_{tn} \qquad (V_{sb} = 0) \tag{3.30}$$

If condition (3.30) held true, the cross-coupled pair voltage doubler would not function at all since positive feedback could not be initiated. Node n_a and node n_b would oscillate in between $-V_D$ and $V_{\text{clk}}\text{CR} - V_D$.

If condition (3.29) held true, it would take a few clock cycles for internal nodes n_a and n_b to reach steady state. Sometimes the initialization process could have performance impact.

Figure 3.23 illustrates a method to enhance the start-up for the cross-coupled pair doubler. Compared with Fig. 3.17, there are two additional NMOS transistors N_3 and N_4 being added. Diode-connected N_3 and N_4 force the initialization of node n_a and node n_b to be at $V_{cc} - V_{tn}$. Some additional parasitic capacitances C_{gs} and C_{db} from startup circuits are added to nodes n_a and n_b.

$$V_{\text{clk}}\text{CR} > 2V_{tn} - V_{\text{cc}} \qquad (V_{sb} = V_{\text{cc}}) \tag{3.31}$$

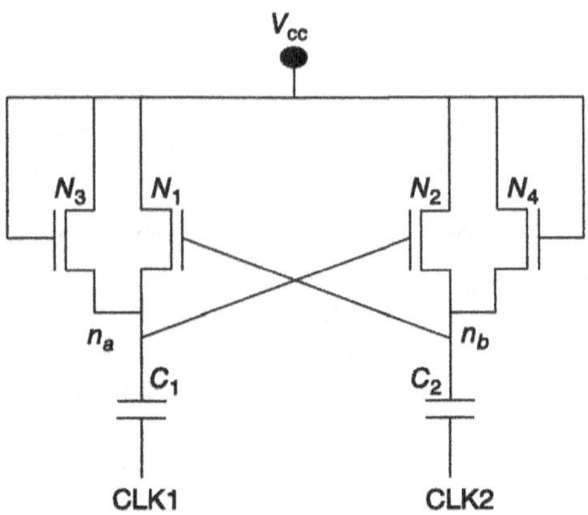

FIGURE 3.23 Start-up circuits with cross-coupled pair voltage doubler.

Compared with Eq. (3.29), Eq. (3.31) shows the reduced minimum criteria to start positive feedback.

3.2.4 Diode Isolation

To utilize cross-coupled pair voltage doubler as the charge pump, diode isolation devices must be added between the internal nodes and output. Figure 3.24 illustrates the configuration of the single-stage, cross-coupled pair voltage doubler charge pump. Diode connected NMOS N_5 is connected between node n_a and V_{out}, and diode-connected NMOS N_6 is connected between node n_b and V_{out}. With this configuration, the single-stage charge pump from the cross-coupled pair could be realized.

3.2.5 Advantages

The single-stage cross-coupled pair voltage doubler has many advantages with relatively simple architecture, such as high power efficiency, low output impedance, high area efficiency, and low output noise.

3.2.5.1 Power Efficiency

As illustrated in Fig. 3.24, there is a simple inherited V_T cancellation scheme built inside the cross-coupled pair voltage doubler. In steady state with the maximum of $2V_{cc}$ on node n_a or n_b, the voltage on the other side of the cross-coupled pair is guaranteed to reach V_{cc} theoretically at the end of the precharge phase. Through positive feedback, the threshold voltage of N_1 or N_2 is effectively canceled in alternating clock phases. There is no addition of devices needed to achieve the V_T cancellation purpose on the input side. Compared with simple diode-isolated input stage, the extra input charge could

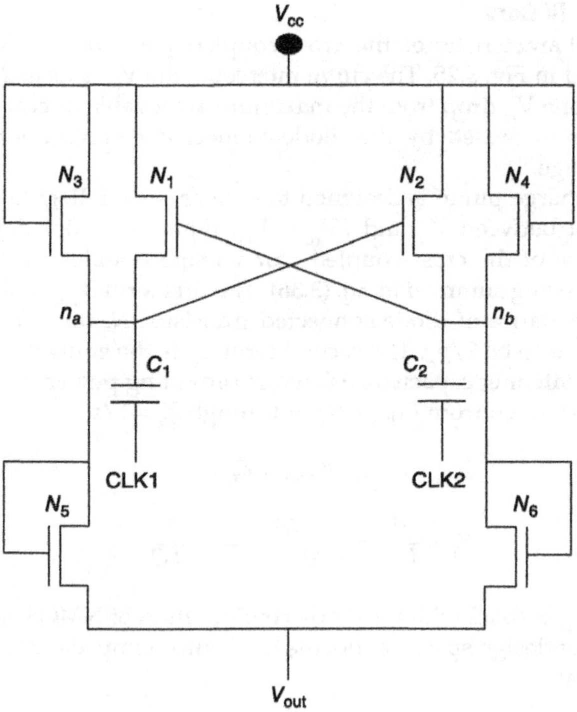

Figure 3.24 Single-stage cross-coupled pair voltage doubler charge pump.

be taken by the pump unit per cycle. The exact amount of extra charge gained is given in Eq. (3.32):

$$\Delta Q = CV_{tn} \tag{3.32}$$

The output current gain I_{gain} could be quantified by Eq. (3.33), and the output power gain P_{gain} could be derived by Eq. (3.34).

$$I_{gain} = \frac{\Delta Q}{T_{period}/2} = \frac{2CV_{tn}}{T_{period}} \tag{3.33}$$

$$P_{gain} = V_{out} I_{gain} = \frac{2\Delta Q V_{out}}{T_{period}} \tag{3.34}$$

$$\text{Power efficiency improvement} = \frac{P_{gain}}{P_{total}} = \frac{V_{tn}}{V_{cc}} \tag{3.35}$$

The power efficiency improvement of the cross-coupled pair voltage doubler over diode-connected charge pump stage could be estimated by using Eq. (3.35).

3.2.5.2 IV Curve

The *IV* characteristic of the cross-coupled pair voltage doubler is illustrated in Fig. 3.25. The curve intercepts the V_{out} axis at $2V_{cc} - V_{tn}$. There is one V_{tn} drop from the maximum achievable internal voltage $2V_{cc}$. This is caused by the diode-connected configuration on the output stage.

The charge pump is designed to deliver power. Regulation level V_{reg} is set between V_{cc} and $2V_{cc} - V_{tn}$. Equation (3.36) derives the impedance of the cross-coupled pair voltage doubler. Two components are being summed in Eq. (3.36). The first term R_{diode} is the equivalent impedance of diode-connected transistor N_5 (or N_6), which is determined to be $1/g_m$. The second term Z_c, is the equivalent impedance of switching capacitance C while providing power to output. It could be derived from Eq. (3.37) or through $Z_c = 1/sC$.

$$R_{eq} = R_{diode} + Z_c \tag{3.36}$$

$$Z_c = \frac{\Delta V}{I_{load}} = \frac{\Delta V}{C\Delta V/T_{period}} = \frac{1}{fC} \tag{3.37}$$

Since R_{diode} is small with the diode configuration of NMOS and Z_c is a function of design specification, the total output impedance R_{eq} is relatively small.

3.2.5.3 Area Efficiency

Area efficiency could be derived from Eq. (3.35). To achieve identical output power specification, the cross-coupled pair voltage doubler can use a smaller capacitor area and supporting circuit area compared with the diode-connected single-stage charge pump. Equation (3.38)

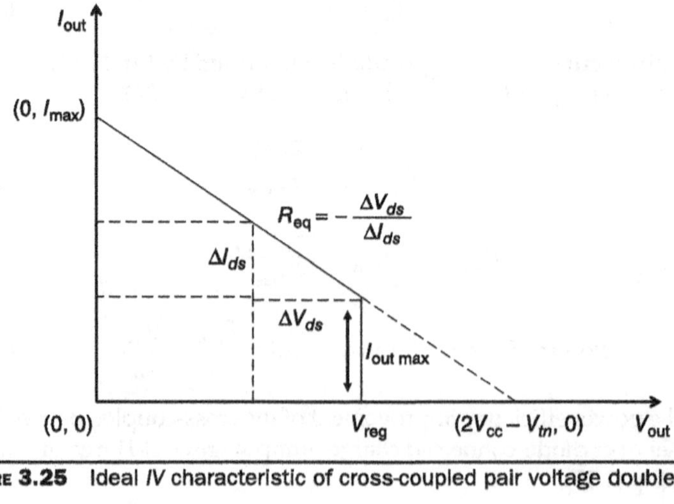

Figure 3.25 Ideal *IV* characteristic of cross-coupled pair voltage doubler.

gives the area aspect ratio with respect to power efficiency between two types of designs.

$$\frac{\text{Area}_2}{\text{Area}_1} = \frac{P_1}{P_2} = \frac{2V_{cc} - V_{tn}}{2V_{cc} - 2V_{tn}} \tag{3.38}$$

3.2.5.4 Output Noise

Since the cross-coupled pair voltage doubler relies on two identical branches to deliver current in two alternating clock phases, the inherited benefit of the architecture determines that the output noise would be suppressed by one-half with the given design.

3.2.6 Architecture Limitations

Although the cross-coupled pair voltage doubler has many advantages over other single-stage charge pump architectures, there is an inherited issue due to the positive feedback which architecture is replying on.

3.2.6.1 Self-loading-Effect due to Load Current

Because of the nature of positive feedback, if the load current were made large enough, the positive feedback of the cross-coupled pair voltage doubler could be turned into negative feedback. The maximum voltage on internal nodes would be reduced from the peak value after the load current exceeded a certain threshold value. If the precharging level were affected by the lowering of boosted voltage on the pass gate connecting between the input supply and internal node, negative feedback would result. Further increase of load current would accelerate the falling of internal voltages until the complete collapse of voltage doubler occurred. Figure 3.25 illustrates the ideal *IV*, and focuses on the impact of maximum achievable output voltage and the change in slope with the cross-coupled pair voltage doubler. Figure 3.26 illustrates the actual *IV* characteristic with the self-loading effect due to the load current.

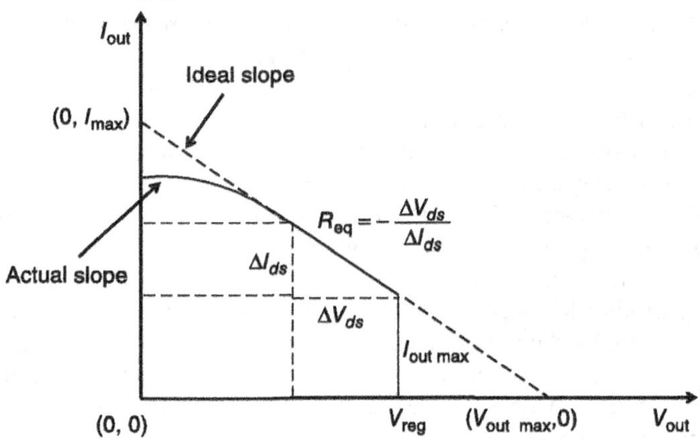

FIGURE 3.26 Actual *IV* characteristic of cross-coupled pair voltage doubler.

With moderate load current on output, the behavior of the *IV* characteristic is expected to follow the linear relationship. The impedance of the cross-coupled pair voltage doubler could be described by Eq. (3.36). As the load current passed a certain threshold, the *IV* curve would deviate dramatically from the ideal slope of the voltage doubler. The impedance of the cross coupled pair voltage doubler would increase exponentially as negative feedback kicked in, and current drivability would fall off exponentially with the lowering of V_{out}. How can one explain this phenomenon?

As shown in Eq. (3.29), which is repeated as Eq. (3.39) for convenience, there is a minimum requirement to sustain positive feedback. In the precharging phase, coupled-up voltage must be able to precharge internal node on the opposite side back to V_{cc}. The coupled voltage depends on the CR and the precharged level from preceding cycle.

$$V_{clk}\mathrm{CR} > V_{tn} \qquad (V_{sb} = V_{cc}) \tag{3.39}$$

Equation (3.39) is for the case when the voltage doubler did not have load current. Equation (3.40) is modified from Eq. (3.39) to address the load current requirement. Voltage V_{ov} is related to the current drivability of the precharge transistor N_1 and N_2 in Fig. 3.24.

$$V_{clk}\mathrm{CR} > V_{tn}(N_1) + V_{ov}(N_1) \tag{3.40}$$

$$I_{ds} = C\frac{\Delta V}{\Delta t} = C\frac{V_{init} + V_{cc}CR - V_{out} - V_{tn}(N_5) - V_{ov}(N_5)}{T_{period}/2} \tag{3.41}$$

$$V_{ov} = \sqrt{\frac{2I_{ds}}{\mu_n C_{ox}\ W/L}} = V_{gs} - V_{tn} \tag{3.42}$$

In the following analysis, only the left half branch of the crossed coupled pair voltage doubler is being analyzed, while the right half circuit could be analyzed in the same fashion. In Eq. (3.40), V_{clk}CR must be greater than the sum of V_{tn} (N_1) and $V_{ov}(N_1)$ to allow proper recovery of N_1 back to V_{cc}. Voltage being coupled up on the gate of N_1 must be able to pass the required charge in the given half clock cycle from the input supply. The average current in the precharge phase through N_1 is equal to the current delivered to the output load through N_5, which is given in Eq. (3.41). As shown in Eq. (3.42), V_{ov} is proportional to the square root of I_{ds} if the gate voltage is fixed. For a fixed W/L ratio, as I_{ds} increases, improving the boosted voltage on associated gates should help to increase V_{ov} for both N_1 and N_5. To a certain extent, as the load current increases, C_1 would not be fully charged up to V_{cc}. In consequence, either the maximum output current under regulation would fall, or V_{out} would drop out from regulation level. If the load current increases further beyond the threshold, precharging and discharging form a negative feedback loop, lowering the precharged voltage and the boosted voltage, and the entire cross-coupled pair voltage doubler would eventually collapse. This phenomenon is described as the self-loading effect of the cross-coupled pair voltage doubler due to load current.

3.2.6.2 Self-loading Effect due to Clock Drivers

Besides the degradation of precharged and boosted voltages on internal nodes which causes negative feedback on the cross-coupled pair voltage doubler, there is another self-loading effect due to clock drives under loading.

The final buffering stage of the clock driver is shown in Fig. 3.27. Either PMOS or NMOS of the clock buffer would be driven by the signal clock to pull up or to pull down the equivalent load C within the given time window.

Figure 3.28 illustrates the equivalent models for pull-up and pull-down paths for the clock buffer. On the left side, R_p represents total equivalent pull-up impedance of PMOS and any associated routings and passing devices, and $1/sC$ is the impedance of coupling capacitance C. On the right side, R_n represents the total pull-down

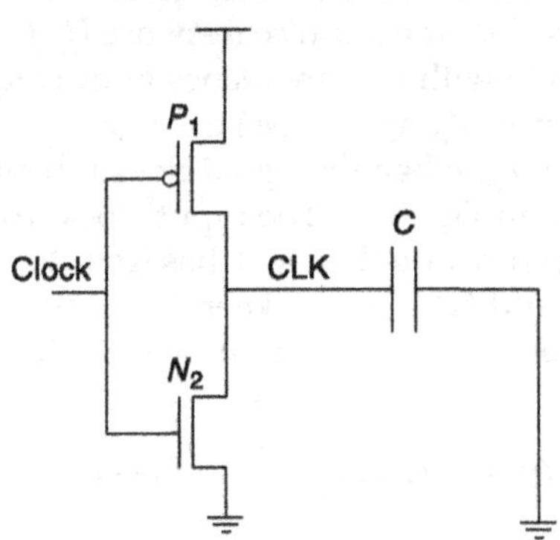

FIGURE 3.27 Clock driver model.

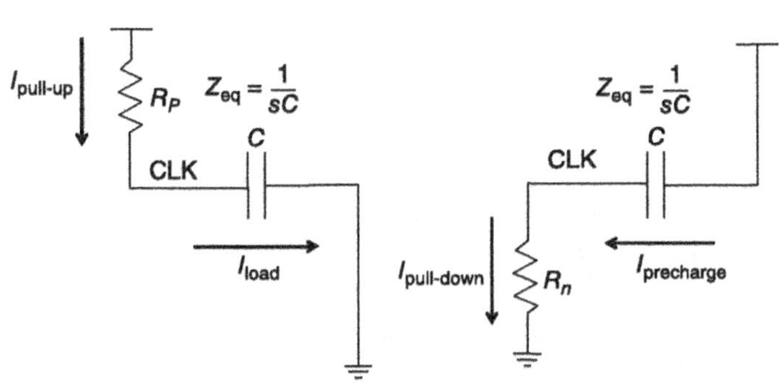

FIGURE 3.28 Equivalent models for pull-up and pull-down paths of clock buffer.

impedance of NMOS and associated routings and passing devices and $1/sC$ is the impedance of coupling capacitance C.

In the simplified model, the equivalent power network impedance is not shown, and the assumption is that it would introduce much less impedance than the buffer. Under steady state, $I_{\text{pull-up}}$ must be equal to $I_{\text{pull-down}}$ to allow the total charge sourced from the preceding stage to flow out to the next stage in alternating pump clock phase. The response of node CLK would follow the first-order linear equation in either the pull-up or pull-down cases as a low-pass network if power were not limited. In design, with low load current, the clock driver can fully charge or discharge node CLK to near final target with tolerable error through linear setting. The time constant $\tau = RC$ would be observed in transient response settling behavior. The clock driver current cannot drop to zero at the end of each clock phase.

$$H_{\text{clk}}(s) = \frac{1/sC}{R_p + 1/sC} = \frac{1}{1 + sR_pC} \tag{3.43}$$

Under this condition, the network forms an impedance divider. Equation (3.43) describes the impulse response $H_{\text{clk}}(s)$ of the network. It is a low-pass function with pole frequency ω_p at $1/R_pC$. The longer the clock period, the closer V_{clk} would be near V_{cc}.

The response of V_{clk} when the signal clock is being driven from V_{cc} to 0 V is illustrated in Fig. 3.29. The input clock drive clock buffer is simplified as a step function that switches from V_{cc} to 0 V at $t = t_0$. For $t < t_1$, PMOS is off and NMOS will keep V_{clk} at 0. At $t > t_0^+$, linear settling behavior is observed on V_{clk}, as shown in Fig. 3.29.

$$V_{\text{clk}}(t) = V_{cc}(1 - \exp^{-\frac{t-t_0}{R_pC}}) \qquad \text{for } t \geq t_0 \tag{3.44}$$

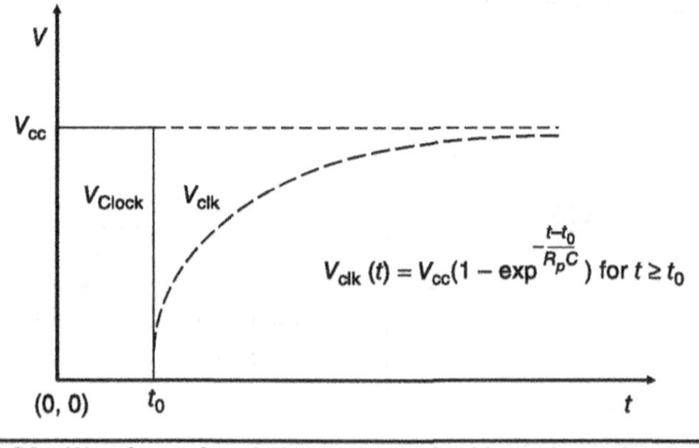

Figure 3.29 Waveform of V_{clk} being pulled high with moderate load.

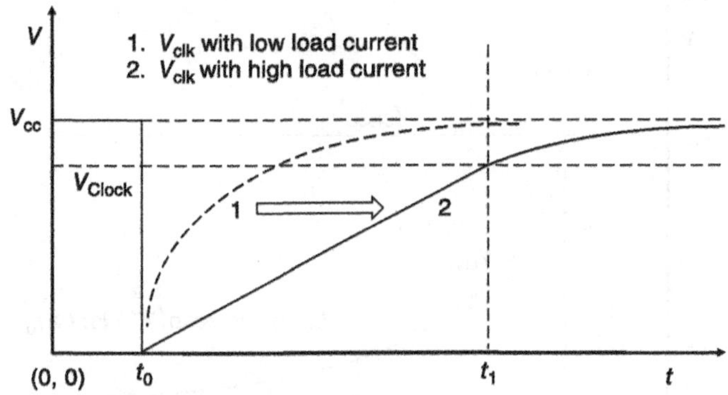

FIGURE 3.30 Waveform of V_{clk} being pulled high with high load current.

Equation (3.44) gives the step response of V_{clk} as a function of time when it was pulled high.

As the load current increases, the limitation of power from the clock driver could constrain the overall response. As illustrated in Fig. 3.30, slewing of V_{clk} highlighted by solid curve 2 would occur in between t_0 and t_1. After $t > t_1$, linear settling would occur. The total response time with addition of slewing is slowed down. To reduce the slew rate, a larger transistor size and good power network are required. Because of the limitation of silicon area and the optimization of power efficiency not at absolute peak load, the buffered pump clocks under high-load condition would have difficulty reaching the maximum target because of the power constraint.

With either linear settling or the slew limitation, buffered clocks could never reach the target in time. The shorter the pump clock period is, the larger the error in clock amplitude from the target. This aspect of self-loading effect from current load is due to the clock driver. It has negative impact on the charge transfer efficiency.

For the pull-down case, the step response of system $H_{clk}(s)$ will follow Eq. (3.45). The node CLK is being pulled down to allow precharge of the coupling capacitance. During the precharge phase, there would be a step function to pull up V_{clk}. Because of the high-pass nature of capacitance and R_n in series, V_{clk} would be coupled up first to V_{cc}. This is indicated by the Zero of $H_{clk}(s)$ in Eq. (3.45). After the initial condition, V_{clk} would follow a first-order linear setting to discharge the bottom plate of capacitance C to 0 V.

$$H_{clk}(s) = \frac{R_n}{R_n + 1/sC} = \frac{SR_nC}{1 + sR_nC} \tag{3.45}$$

$$V_{clk}(t) = V_{cc}(\exp^{-\frac{t-t_0}{R_nC}}) \qquad \text{for } t \geq t_0 \tag{3.46}$$

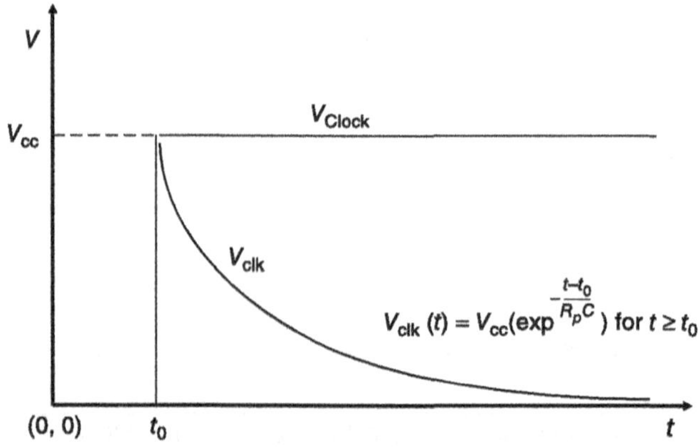

FIGURE 3.31 Waveform of V_{clk} being pulled low with moderate load current.

Figure 3.31 illustrates the time-domain waveform of V_{clk} base on Eq. (3.45). Equation (3.46) gives the response of V_{clk} over time. Since the pull-down current would not settle down to 0 mA, there is always the residue potential left on V_{clk} instead of being pulled down to 0 V.

Figure 3.32 illustrates the response of V_{clk} under heavy load conditions with highlighted curve 2. Because of the slew rate limitation, the response of V_{clk} being pulled low is slowed down from curve 1 to curve 2. The linear settling happens only after $t > t_1$ until power is no longer a limitation.

Similar to the pull-up case, with linear settling or with slew limitation, the response of V_{clk} would deviate from ideal behavior because of the load current. The self-loading effect on the pull-down path of the clock driver would impact the response of voltage doubler.

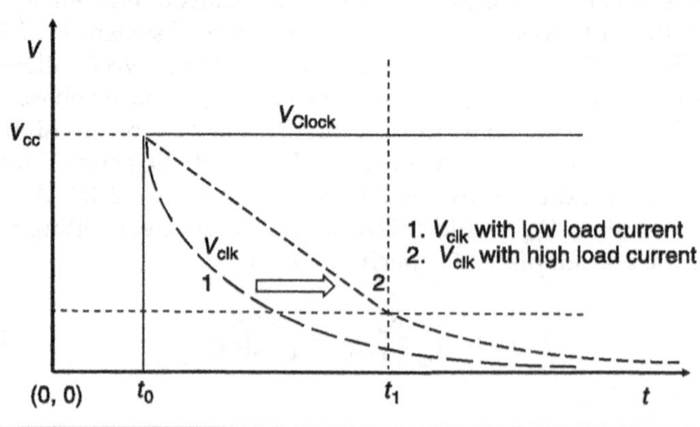

FIGURE 3.32 Waveform of V_{clk} being pulled low with high load current.

3.3 Cross-Coupled Pair Voltage Doubler: Parallel Stage V_T Cancellation on Output Stages

In Sec. 3.2, the simple cross-coupled pair voltage doubler is examined in detail. On one hand, there are a lot of advantages associated with the given architecture, such as simple V_T cancellation on input transistors without extra circuits, better power efficiency and area efficiency, low output impedance, and minimized output voltage noise. On the other hand, there are a few disadvantages. There is no V_T cancellation on output isolation transistors, and there are self-loading effects due to load and clock drivers under heavy load current. In the worse case, positive feedback of the cross-coupled pair voltage doubler could be turned into negative feedback and collapse the pump supply under heavy load. Conventional V_T cancellation schemes could be applied to output isolation transistors. The design could be complex, and it would not be immune from self-loading effects since internal voltages would be impacted by load.

How is one to overcome the disadvantages of the basic cross-coupled pair voltage doubler? A new architecture is needed to address the concerns. Figure 3.33 illustrates a new architecture. It is a cross-coupled pair voltage doubler with parallel stage V_T cancellation on the output stages.[2]

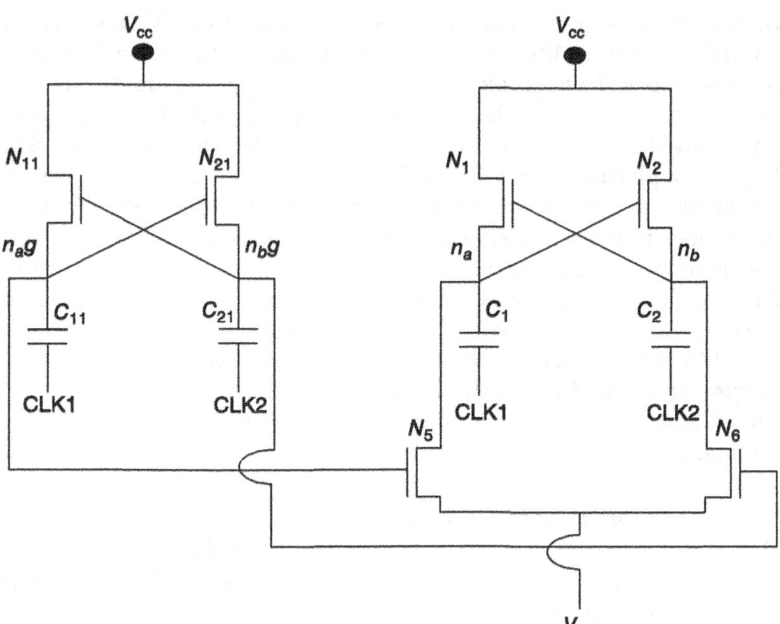

FIGURE 3.33 Cross-coupled pair voltage doubler with parallel stage V_T cancellation on the output stages.

Two identical cross-coupled pair voltage doublers are placed side by side. Except for the size difference, the circuit structures are identical. For isolation transistors N_5 and N_6, instead of connecting as diodes, the gate of N_5 is connected to node $n_a g$ from the parallel stage, and the gate of N_6 is connected to node $n_b g$ from the parallel stage. How can this architecture perform better than the basic cross-coupled pair voltage doubler discussed in Sec. 3.2?

3.3.1 Advantages

Compared with the original design, the newly introduced parallel stage is used for V_T cancellation of isolation transistors in the output stages. How does this V_T cancellation scheme perform against other types of V_T cancellation schemes?

As described in the literature, the most commonly adopted V_T cancellation scheme is probably based upon the four-phase charge pump. Four different clock phases are required to perform the V_T cancellation function. Two main pump clocks overlap in high phases for the purpose of precharge and discharge of gates of passing elements. Two auxiliary pump clocks have high clock phases embedded in the main pump clocks to couple up the gate of passing elements and transfer the charge fully without V_T limitation. As shown in the analysis of charge pump circuit design,[3] four pump clocks come from different clock drivers. To maintain the phase margins of the clocks in the worst PVT corners, the design margins could be wasted in other PVT corners. In addition, owing to the mismatches of layout, routings, loading, and different clock driver designs, more timing margins must be kept in the design to cover all the possible variations. Both the design and layout are challenging. In terms of charge transfer, the auxiliary clock's high clock phase cannot take the 50% duty cycle of the pump clock. If a 1–ns time margin is needed between the main pump clock and auxiliary clock in terms of rising and falling edges, and if the pump clock was running at 20 Mhz, or a 50-ns clock period, then 2 ns out of the 25-ns half clock period was not used by the design to deliver power to the output. This design will incur additional output noise.

With the goal to simplify the design and layout effort and to maximize of the duty cycle of charge transfer, the parallel stage V_T cancellation scheme on the output stages of the cross-coupled pair voltage doubler is developed.

3.3.1.1 V_T Cancellation on Output Stage

Why choosing a parallel stage of the cross-coupled pair voltage doubler to drive the gate of NMOS N_5 and N_6? Let's revisit the discussion of V_T cancellation from Sec. 3.1.3.

In Fig. 3.34, the *IV* curve of the diode-connected NMOS N_5 is highlighted as a solid line. The family of *IV* curves of NMOS was plotted for reference. For the purpose of charge transfer, the characteristic similar to the ideal *IV* curve of NMOS is strongly preferred.

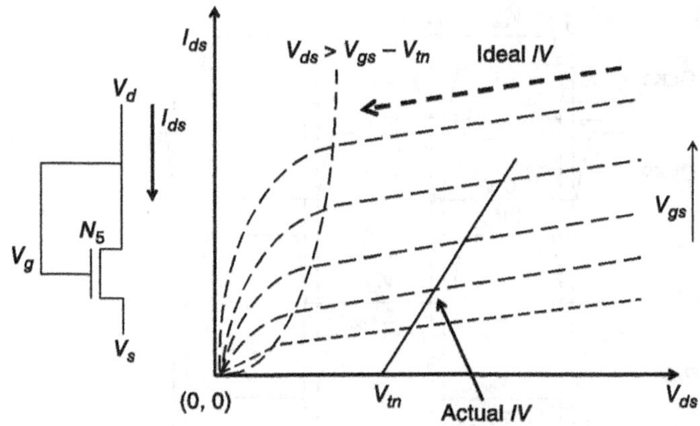

FIGURE 3.34 Limitation of NMOS diode-connected device.

The *IV* curve of diode-connected N_5 has weak current capability as both V_{gs} and V_{ds} are reduced. As V_{gs} drops to V_{tn} or below, the transistor will be cut off.

If the gate of the NMOS transistor could be held steady during charge transfer as Vds reduces, then NMOS would act as a current source as long as $V_{ds} \geq V_{gs} - V_t$. As V_{ds} is further reduced by charge transfer from drain to source, the NMOS transistor enters into the linear region until V_{ds} drops to 0. This is the ideal characteristic that isolation devices should hold in the charge transfer phase.

With the introduction of the parallel stage, there are two identical stages: One is the main cross-coupled pair voltage doubler to deliver power to the output, and one is the parallel stage to drive the gates of passing elements in the main pump stage. Since the gate stage drives only the gates of N_5 and N_6, the loading to the gate stage is much smaller than that of the main stage. All the transistors and coupling capacitance could be down sized easily to occupy a much smaller area and consume less power.

Figure 3.35 illustrates the clock waveforms of parallel stage V_T cancellation on the output stage without current load. CLK1 and CLK2 are pump clocks. Nodes n_ag and n_bg are located in the parallel gate stage. Nodes n_a and n_b are located in main stage. Without load, it is easy demonstrate the concept of the parallel stage V_T cancellation scheme.

Pump clocks CLK1 and CLK2 are nonoverlapping clocks with Δt separation between high phases. They switch between 0 V and V_{cc}. With capacitive coupling, the internal nodes of both the gate stage and the main stage operate in phase with CLK1 and CLK2. Similarly, n_a and n_ag and n_b and n_bg would have identical clock phases too. In steady state they switch between V_{cc} and $2V_{cc}$ without load at output. Under the equilibrium condition without load, the parallel stage V_T cancellation scheme would have the same biases on all devices as those in the normal cross-coupled pair voltage doubler.

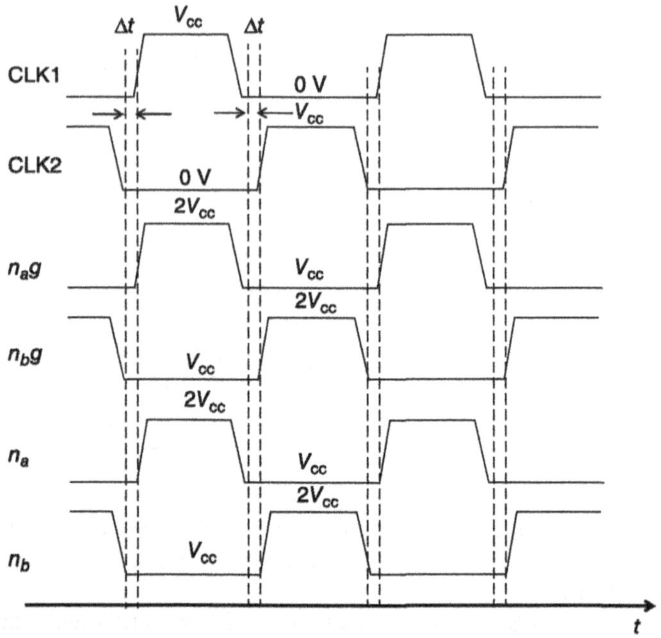

FIGURE 3.35 Waveforms of parallel stage V_T cancellation on output stage without current load.

Figure 3.36 illustrates the waveforms of internal nodes of parallel stage V_T cancellation on the output stage with current load. In alternating clock phases, node n_a and n_b are coupled up and then discharged to output. Within the same clock phase, as V_{ds} across NMOS N_5 or N_6 drops, node n_ag or node n_bg stays at the $2V_{cc}$ level. The load current or charge transfer does not influence the gate voltage of N_5 or N_6. In the precharge phase to recover node n_a or n_b back to V_{cc}, node n_ag or n_bg would stay at V_{cc}. NMOS transistor N_5 or N_6 acts as diode-connected device to prevent charge to leaks backward.

Figure 3.37 illustrates the overlapped waveforms of internal nodes with and without current load. While CLK1 switches from 0 V to V_{cc}, node n_ag and n_a are boosted up to $2V_{cc}$. Charge is transferred from node n_a to output, and V_{ds} of NMOS N_5 reduces. NMOS N_5 operates in the saturation region at the beginning. As V_{ds} drops below $V_{gs} - V_{tn}$, the device enters into the linear region. The maximum speed of transition is realized, and the maximum amount of charge is transferred. When CLK1 switches from V_{cc} to 0 V, node n_ag is coupled back to V_{cc}. Node n_a would be coupled below V_{cc}, and then back to V_{cc}. The level of node n_a settled in transition is a function of load current. Momentarily there would be reverse leakage from the output back to node n_a depending on the voltage level of node n_a and the threshold of NMOS N_5. Since node n_b would be coupled up after Δt following the

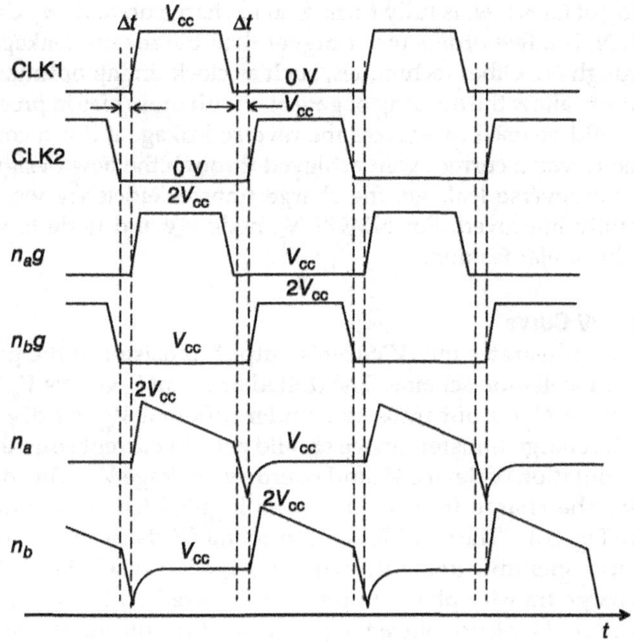

FIGURE 3.36 Waveforms of parallel stage V_T cancellation on output stage with current load.

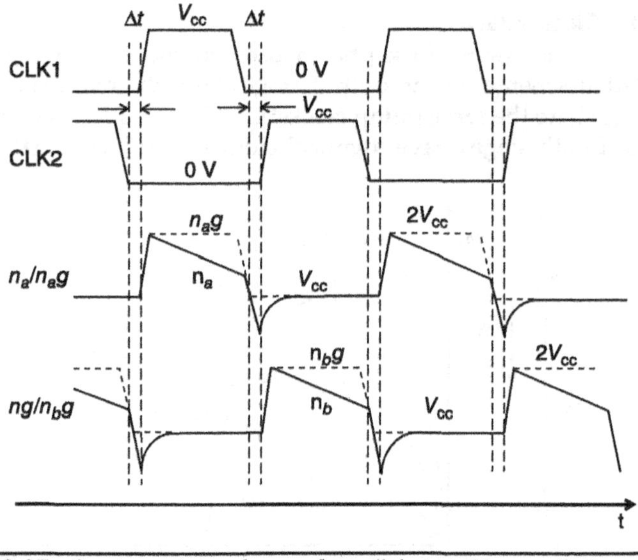

FIGURE 3.37 Overlapped waveforms of parallel stage V_T cancellation on output stage with current load.

switching of CLK1, N_1 is fully turned on to charge up node n_a. Current through N_1 is a few orders order higher than the reverse leakage current through N_5. Other techniques, such as clock timing optimization, or circuits to allow the tracking of gate bias with drain bias in precharge phase[4] could be used to address the reverse leakage if it is a concern. Since the forward current gain achieved through the new design outweighs the reverse leakage, the charge transfer efficiency would be significantly improved. For NMOS N_6, node n_bg and node n_b would operate in similar fashion.

3.3.1.2 IV Curve

Figure 3.38 illustrates the *IV* curve's output transistor in the parallel stage V_T cancellation scheme. The dotted lines are I_{ds} versus V_{ds} of the NMOS N_5 or N_6 output transistor under different V_{gs}. As discussed earlier, the charge transfer device should act as a current source without the limitation of device V_T and overdrive voltage V_{ov}. The ideal *IV* curve for the charge transfer device is highlighted by the dashed arrow in Fig. 3.38. The solid line is the actual *IV* characteristic of output devices operating under the parallel stage V_T cancellation scheme. In the charge transfer phase, node n_ag is boosted up to $2V_{cc}$ and will not be affected by load. Since the gate bias of the output transistor is not changing in the charge transfer phase, NMOS transistor N_5 or N_6 acts as an ideal current source for the majority of the operation period before it enters the linear region. With the new design, there is no V_T limitation as diode-connected transistor.

3.3.1.3 Clock Phases

Figure 3.39 illustrates clock phases used in the parallel stage V_T cancellation scheme on the output stage. Both the main stage and gate stage share the same pump clocks: CLK1 and CLK2. Is there any concern if both stages have identical clock phases? The analysis of

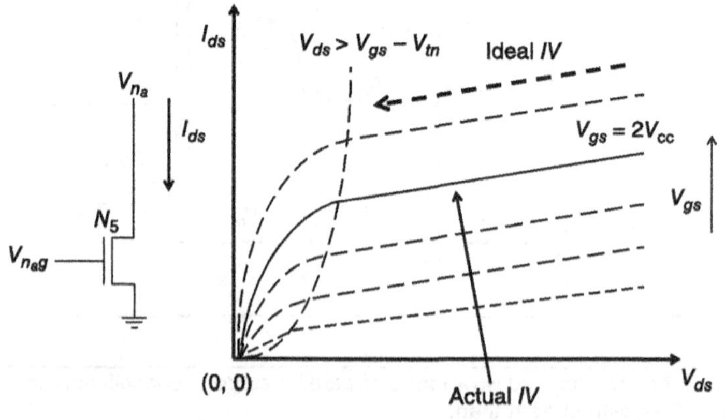

FIGURE 3.38 The *IV* curve of the output transistor in parallel stage V_T cancellation scheme.

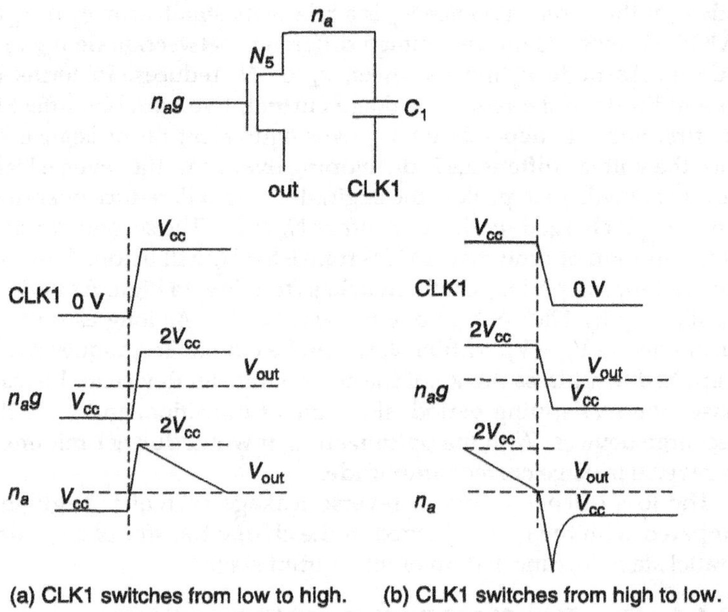

FIGURE 3.39 Clock phases in parallel stage V_T cancellation.

clock phases could be divided into two parts: high switching and low switching transitions. Figure 3.39 illustrates the corresponding waveforms of all internal biases related to NMOS N_5.

In Fig. 3.39*a*, when CLK1 switches from low to high in the charge transfer phase, both node n_a and node n_ag are coupled up by CLK1 at the same rate through AC coupling. Node N_5 is in the cutoff stage until Vn_ag is V_{tn} above V_{out}. Even if there was any mismatch or phase shift, the gate and main stages track one another and the charge could still be transferred from node n_a to node out. Before CLK1 makes the high-to-low transition, N_5 operates in the linear region since node n_a and node out are at nearly equal potential.

In Fig. 3.39*b*, CLK1 switches from high to low in the precharging phase. NMOS N_5 should be designed to minimize the reverse leakage if it occurred with high current load.

Since node n_a and node n_ag switch from high to low in phase with CLK1, and node out is at higher potential with respect to node n_a. NMOS N_5 operates in the linear region as given by.

$$I_{ds} = \mu_n C_{ox} \frac{W}{L}[(V_{gs} - V_t)V_{ds} - \tfrac{1}{2}V_{ds}^2] \tag{3.47}$$

$$g_{ds} = \frac{1}{r_{ds}} = \mu_n C_{ox} \frac{W}{L}(V_{gs} - V_t) \tag{3.48}$$

In Eq. (3.47), node out is the drain, node n_a is the source, and node n_ag is the gate. Node out is at higher potential and will leak back into node n_a at this moment. Since V_{ds} is a relatively small term, g_{ds} or r_{ds} of NMOS N_5 depends on the voltage difference between node n_ag and node n_a. As node n_a moves lower, r_{ds} of N_5 reduces. In terms of the amplitude of the reverse leakage current, two variables affect it. The first variable depends on V_{ds} over equivalent on resistance r_{ds}. Since the voltage difference is developing over time, the reverse leakage current will go to peak at the beginning and will reduce over time as node n_a is charged up through either N_5 or N_1. The second variable is the duration of time that NMOS transistor N_5 will be on. After the nonoverlapping period, CLK2 switches from low to high. As node n_a is pulled up by NMOS N_1, node n_ag stays at V_{cc}. As long as node n_a reaches above $V_{cc} - V_t$, NMOS N_5 would be cut off. Techniques could be applied to address the V_{gs} of the charge transfer device in the reset phase, nonoverlapping period, slew rate of transition, and g_m of the precharge devices. With the optimization, it would further minimize the reverse leakage current amplitude.

The loss of charge due to reverse leakage current is negligible compared with the gain achieved in the charge transfer phase using parallel stage V_T cancellation on the output stage.

3.3.1.4 Floor Plan, Signal Routing, and Matching

Physical design is one of most important factors that could influence the outcome of pump design. Architecture can help to ease the design constraints on the layout.

Parallel stage V_T cancellation use two parallel stages with identical structures. In terms of floor planning, two stages could be easily laid out side by side with identical placement of devices. Signals and the power plan could be duplicated with scaling to enable the best matching.

The timing margin is one of the most critical design metrics that determines the charge pump efficiency. Complex architectures with multi phase clocks would have tight constraints in circuit design and layout. To ensure the timing margin in worst corners, variations due to devices, signal paths, and mismatches must be budgeted into the design margin. There would be loss of efficiency in other PVT corners.

For parallel stage V_T cancellation architecture, since both the gate stage and the main stage are matched in design and layout, with common pump clocks driving both stages, the components related to variations and mismatches of pump clocks are significantly reduced. The architecture itself enables the simplicity of design and layout, which in turn minimized the timing margin required in design.

3.3.1.5 Power Efficiency

In terms of power efficiency, it is best to analyze in terms of Thevenin's equivalent resistance of the charge pump. Equation (3.49) shows there are two components in output impedance R_{eq}: one is related to the on resistance of isolation transistor N_5 and N_6, and the other one is related to the switch capacitance C_1 and C_2. The assumption is that capacitors C_1 and C_2 could be fully precharged to V_{cc} during the corresponding precharging phase of each capacitor.

$$R_{eq} = R_{on} + \frac{1}{sC} \tag{3.49}$$

$$R_{on} = \frac{1}{\mu_n C_{ox}\, W/L(2V_{cc} - V_{out} - V_t)} \tag{3.50}$$

With this new scheme, the impedance related of C_1 and C_2 are not changed. Here $1/sC$ is the component that would not be altered from the basic cross-coupled pair voltage doubler. The component that has been improved significantly is R_{on}. Equation (3.50) illustrates R_{on} with passing elements in the new scheme. Since the gate bias is maintained at $2V_{cc}$, the on resistance of the passing device is significantly lower than on resistance diode-connected NMOS.

Figure 3.40 illustrates how the *IV* curve could be improved with parallel stage V_T cancellation on the output stage compared with the basic cross-coupled pair voltage doubler. In the figure, two *IV* curves are shown. The lower curve is the *IV* curve of the basic cross-coupled pair voltage doubler. It intercepts the V_{out} axis at $2V_{CC} - V_{tn}$, and it intercepts the I_{out} axis at I_{max1}. The target regulation level is at V_{reg}. The total available current to sustain regulation is I_{reg}(old). The inverse of slope of *IV* curve is the equivalent output impedance R_{eq1} of the original design without V_T cancellation on the output stage. The upper curve is the *IV* curve of the cross-coupled pair voltage doubler with

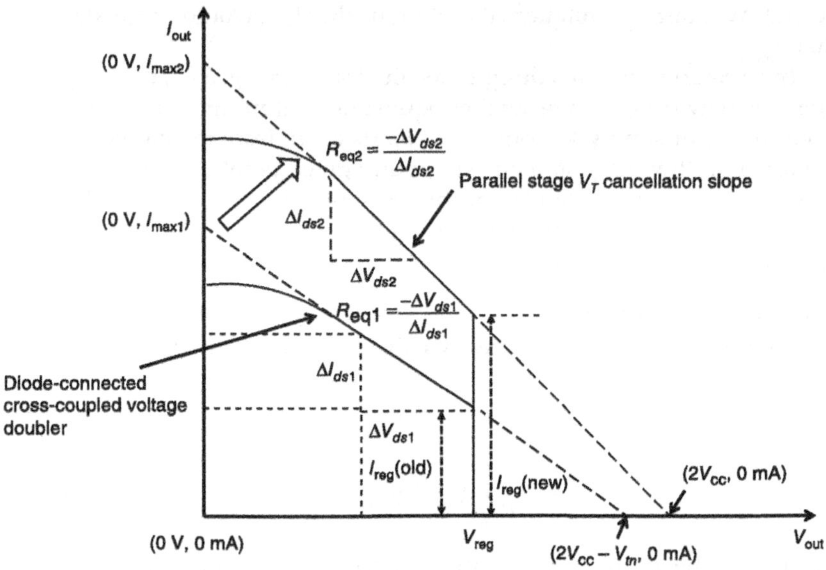

FIGURE 3.40 *IV* characteristic improvement of parallel stage V_T cancellations on output stage of cross-coupled pair voltage doubler.

parallel stage V_T cancellation on the output stage. It intercepts the V_{out} axis at $2V_{cc}$, and it intercepts the I_{out} axis at I_{max2}. The target regulation level is same as in the first design, which is at V_{reg}. The total available current to sustain regulation is I_{reg}(new). The new design has an equivalent output impedance R_{eq2}. The shallower slope of the *IV* curve of the old design indicates the relatively high equivalent output impedance of design, which is not optimum as a power supply. The new design has steeper slope, which indicates lower output impedance compared with the old design.

$$I_{max1} = \frac{\partial Q_1}{\partial t} = \frac{2C(V_{cc} - V_{tn})}{t_{period}} \tag{3.51}$$

$$I_{max2} = \frac{\partial Q_2}{\partial t} = \frac{2CV_{cc}}{t_{period}} \tag{3.52}$$

In terms of the maximum current that a cross-coupled pair voltage doubler could deliver, it is easy to analyze in terms of the total charge that would be transferred per half pump clock cycle. In a basic cross-coupled pair voltage doubler, the maximum amount of charge stored before coupling is the product of C and the DC average potential voltage. The maximum charge that could be delivered to the output with pump clock amplitude V_{cc} is given by Eq. (3.51). There is a limitation of V_{tn} with the body bias effect limiting the total charge transfer. In terms of the parallel stage V_T cancellation architecture, Eq. (3.52) shows the maximum current that could be delivered to the output. Without V_T limitation, it is obvious that I_{max2} would be greater than I_{max1}.

In comparing the two designs, as illustrated in Fig. 3.40, not only does the new design have smaller equivalent output impedance R_{eq} from the input supply V_{cc}, but also it shifts away from the old curve to gain in both voltage and current at any given point. At the target regulation level, the amount of output current to maintain the regulation level is I_{reg}(new) in the new design, which is much larger than I_{reg}(old) of the old design.

3.3.1.6 Area Efficiency

Power efficiency can be converted easily into area efficiency.

$$R_{eq} = -\frac{V_{out} - V_{max}}{I_{out}} \tag{3.53}$$

$$I_{reg} = -\frac{1}{R_{eq}}(V_{reg} - V_{max}) \tag{3.54}$$

Assuming the output specification is fixed for all designs, with known impedance R_{eq} from the design, the I_{ds} versus V_{out} function could be constructed as given by Eq. (3.53). Since the target regulation voltage is V_{reg}, the maximum regulation current I_{reg} at a given regulation level could be derived from Eq. (3.54). For pump design to meet the output specification with fixed architecture, the product of total capacitance area and output current capability is a constant, as shown in Eq. (3.55). The area efficiency between the two designs could be estimated easily by using Eq. (3.56).

$$\frac{\text{Area(old)}}{\text{Area(new)}} = \frac{I_{reg}(\text{new})}{I_{reg}(\text{old})} \tag{3.55}$$

$$\text{Area efficiency} = \frac{\text{area(old)} - \text{area(new)}}{\text{area(old)}} = 1 - \frac{I_{reg}(\text{old})}{I_{reg}(\text{new})} \tag{3.56}$$

3.3.2 Architecture Limitations

As derived earlier in this chapter, the cross-coupled pair voltage doubler with parallel stage V_T cancellation on the output stage architecture has many advantages for designers. However, there are a few limitations that design could not address. All are related to self-loading effects due to load current.

3.3.2.1 Self-loading Impact in Precharging Phase

As shown in Fig. 3.40, both old and new architectures are suffering from self-loading effects in the precharge phase in the main stages with large load current. This phenomenon will always occur at the start-up state of charge pump operation with the load connected, or if huge capacitive load is connected to the output of charge pump, or if a peak current load dump occurs.

At the start-up stage, V_{out} is at very low potential and could sink very high current from the charge pump. It will take time to move the *IV* curve out of this degenerated region before the pump could follow the normal slope of the *IV* curve as V_{out} rises. Inefficiency in V_T cancellation on the input stages due to non-steady-state internal voltages slows down the transient settling time.

The same argument could be applied to the scenario with huge capacitive load switching or peak current load dump on the output of the charge pump in operation. The settling or recovery time is one of the important design specifications. The timings could be impacted significantly since the charge pump output power capability is deviating from the ideal level.

3.3.2.2 Self-loading Effect due to Sharing of Pump Clock Drivers

There is second self-loading effect due to the sharing of clock drivers between the main stage and gate stage under current load.

Figure 3.41 illustrates the schematic of the sharing of pump clock drivers between the gate stage and the main stage in parallel stage V_T cancellation on output stages, shown in Fig. 3.33. The main stage has

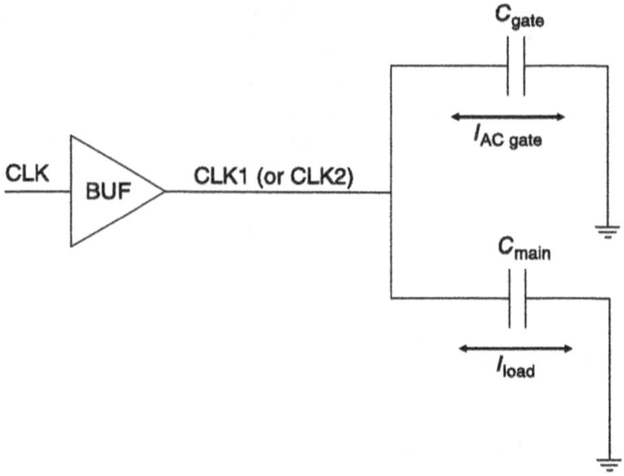

Figure 3.41 Sharing of pump clock drivers between gate stage and main stage.

load current I_{load} that could follow bidirectionally. The gate stage has AC load current I_{AC} gate. Since there is a common node CLK1 (or CLK2) that is connected between two stages through capacitors C_{main} and C_{gate}, current I_{load} through C_{main} could impact the potential or slew rate on node CLK1 (or CLK2). In consequence the change of voltage of CLK1 (or CLK2) would feed into the gate stage and impact node n_ag or node n_bg in V_T cancellation. This is the self-loading effect due to clock driver sharing.

Figure 3.42 illustrates the impact from the load current on the internal node of the gate stages. Current I_{load} or $I_{precharge}$ always dominates I_{AC}. In the pull up case, CLK is being driven high to deliver I_{load}. With V_T cancellation on the output stage, the on resistance of the output

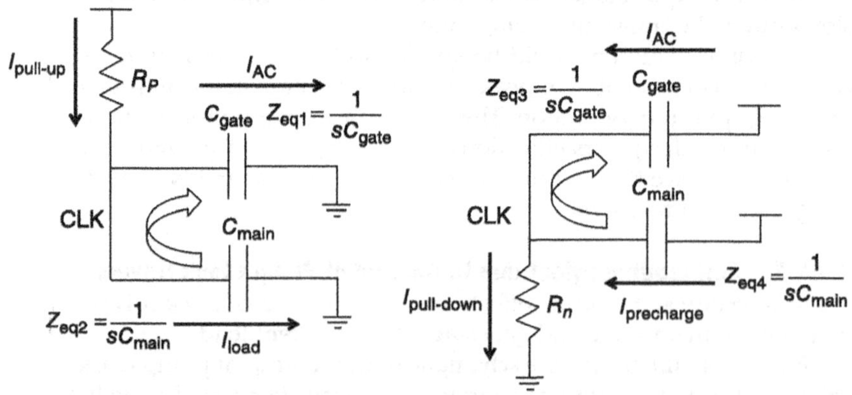

Figure 3.42 Equivalent models of self-loading effects by sharing clocks.

transistor is low. Node n_a or node n_b would quickly equalize with the output and voltage drop. Since the voltage across capacitance C_{main} could not be changed instantaneously, CLK would be coupled lower by node n_a or node n_b. Eventually CLK would recover back to V_{cc} by $I_{pull\ up}$. While node CLK is coupled lower, noise migrates into node n_ag or node n_bg through C_{gate}. The time constant for node CLK recovery is about $R_p(C_{main} + C_{gate})$. Eventually CLK would recover back to V_{cc} within the given pump clock cycle. However, during this recovery phase either node n_ag or node n_bg is not at its targeted level of $2V_{cc}$. This nonideal bias will impact both the precharging of the internal node, and the charge transfer to output. In the pull-down case, similar arguments could be applied.

When the load current is large enough, the *IV* curve will further deviate from the target because of load noise interference through common clock drivers. Figure 3.43 illustrates how self-loading effect due to shared clock drivers impacts the *IV* curve. Curve 1 is the *IV* curve of ideal design. Curve 2 is the *IV* curve with the self-loading effect due to load current. Curve 3 is the *IV* curve with self-loading effects due to both the load current and the shared clock drivers. As the load current decreases, all cases will converge to the same interception point on the V_{out} axis. At higher load current, case 3 would deviate further from case 2. Under the same output voltage, case 3 delivers the least amount of power in the high and median load current ranges.

The self-loading effect due to the sharing of clock drivers can be easily resolved with split clock driver design. Figure 3.44 illustrates that CLK1G and CLK1M are buffered from CLK1, and CLK2G and

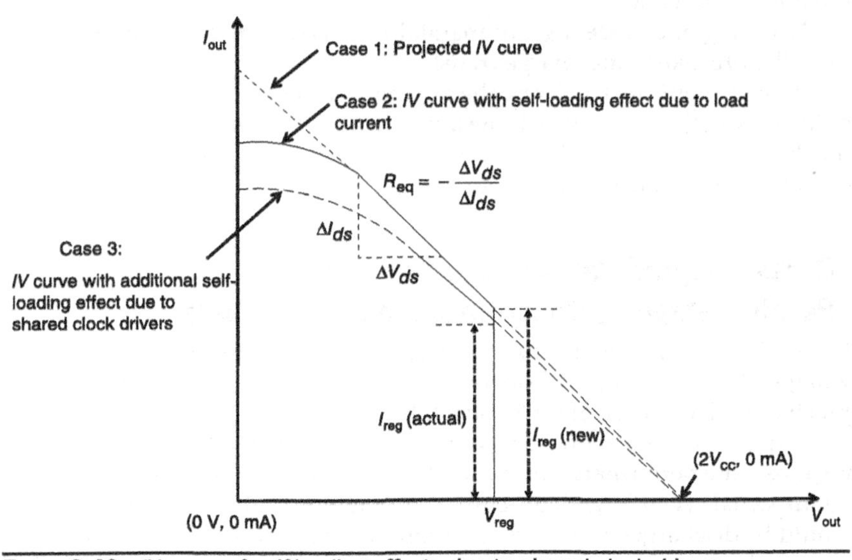

FIGURE 3.43 *IV* curve of self-loading effects due to shared clock drivers.

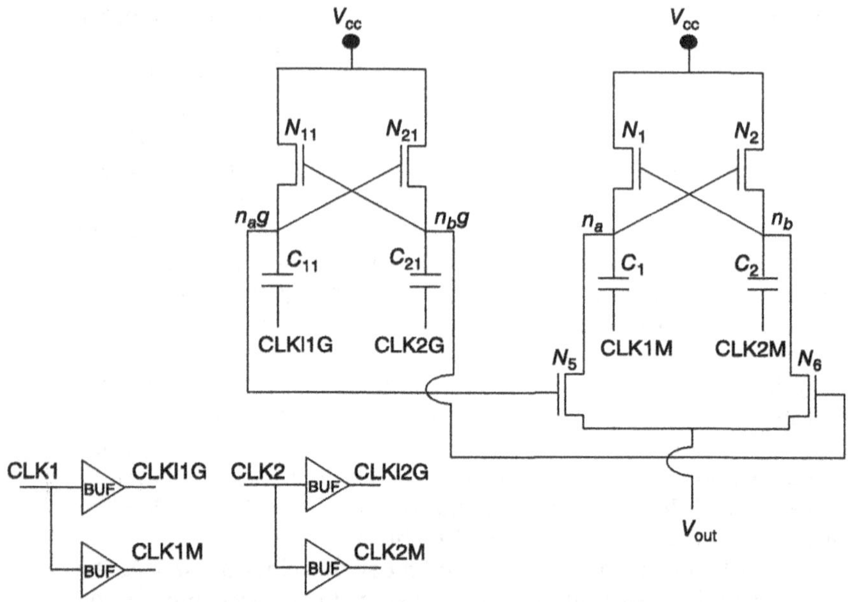

Figure 3.44 Parallel stage V_T cancellation on output stage with split clock drivers.

CLK2M are buffered from CLK2. If the load current is large, then noise on CLK1M is isolated from CLK1G and noise on CLK2M is isolated from CLK2G. Through the isolation of buffers and noising filtering from power supplies, gate stage potentials are immune from the noise from load.

Node n_ag and node n_bg can maintain at maximum $2V_{cc}$ to maximize the precharge and charge transfer.

In using split clock drivers, the assumption is that the impedance of the power network is low. If the layout was not taken care properly for the local power supply network, split clock drivers are still susceptible to noise interference from load.

3.4 Cross-Coupled Pair Voltage Doubler: Parallel Stage V_T Cancellation on Input Stage

In Sec. 3.2.6.2, one architecture limitation of the cross-coupled pair voltage doubler is the self-loading effect due to load current in precharge phase. If positive feedback turned into negative feedback with high load current, the pump could collapse. In transient response, the performance of the start-up or recovery response could be impacted. With large load current, the internal coupled voltages would be discharged too quick. The consequence is the total integration of charge transferred in precharging phase would be reduced.

Reduced charge per clock cycle translates to reduced output current, or less output power capability. This problem would become much more severe with the scaling down trend of chip power supplies, combined with the high threshold voltage of input transistors.

Is there a way to resolve the self-loading effect due to large load current in the precharge phase? Figure 3.45 illustrates a new architecture to implement the cross-coupled pair voltage doubler with parallel stage V_T cancellation on the input stage. In Fig. 3.45, there are two identical stages with different sizing: the main stage and the gate stage. The gate stage is the cross-coupled pair voltage doubler. The main stage is like a cross-coupled pair except for the connections of the gates of NMOS N_1 and N_2. The gate of N_1 is connected to node n_bg rather than node n_b, and the gate of N_2 is connected to node n_ag rather than node n_a. CLK1G and CLK1M are buffered from CLK1, and they are in phase. CLK2G and CLK2M are buffered from CLK2, and they are in phase too. From the clock phase point of view, the main stage indeed operates as a cross-coupled pair voltage doubler.

3.4.1 Advantages

As discussed in Sec. 3.2.6.2, the self-loading effect does impact the performance, timing, and functionality of design. The total charge

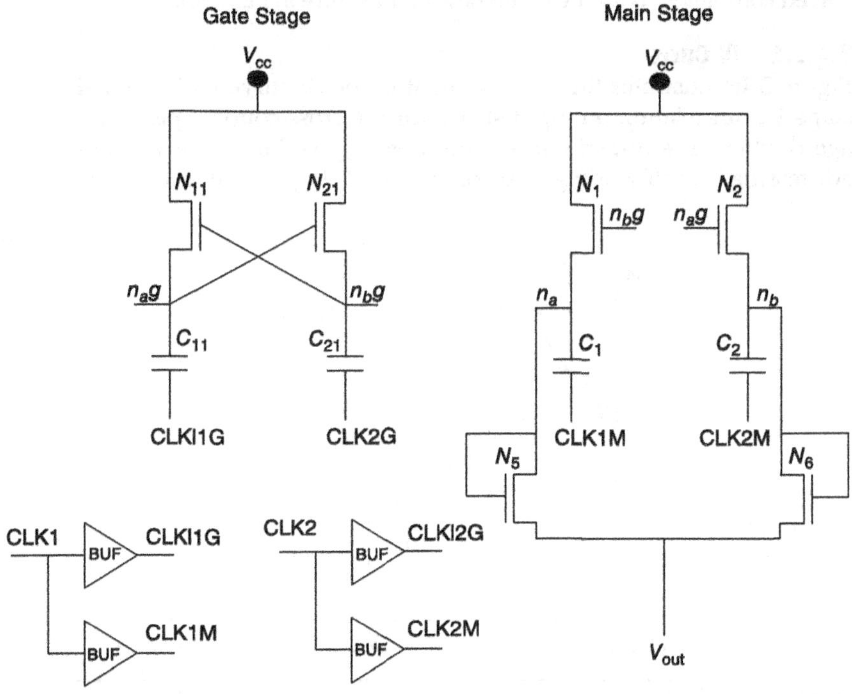

FIGURE 3.45 Parallel stage V_T cancellation on input stages.

stored on capacitor C_1 or C_2 in the precharge phase is the integration of current through NMOS N_1 or N_2 over a half clock cycle. Equation (3.57) gives the estimate of total integrated charge over half a clock cycle, where Q is the total integrated charge stored on capacitor C_1 or C_2 within a half clock cycle. The conduction current of NMOS N_1 or N_2 is a function of V_{gs} across the transistor. The gate bias is the coupled potential. It decays as a function of time and load current. As the load current increases, the boosted gate voltage decays faster. Once the load current exceeds a certain threshold, the node being driven for precharging will not reach V_{cc}. Negative feedback occurs and the output current capability quickly degrades if the load current continues to increase.

$$Q = \int_0^{t_{cycle}/2} I_{ds}\{V_{gs}[I_{load}(t)]\}\, dt \tag{3.57}$$

The new architecture illustrated in Fig. 3.45 decouples completely the gate biases of NMOS N_1 and N_2 from the main stage output load. The load on node $n_a g$ and node $n_b g$ is the fixed capacitive load to gate stage. The gate voltage of N_1 or N_2 is a fixed potential at $2V_{cc}$ in the precharge phase. Without the influence of the output load, the fixed potential on the gate would allow maximum drivability of N_1 or N_2 in the precharge phase. It is clear that the self-loading effect due to the current load is solved completely with the given new architecture in Fig. 3.45.

3.4.1.1 IV Curve

Figure 3.46 illustrates the improvement of the *IV* curve with parallel stage V_T cancellation on input stages for the cross-coupled pair voltage doubler. The dotted curve with lower I_{out} is the design without addressing the self-loading effect on the input stage. Positive feedback

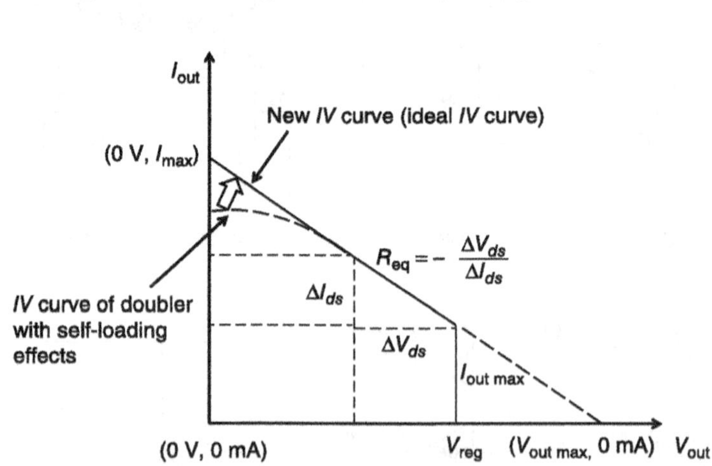

Figure 3.46 The *IV* of curve cross-coupled pair voltage doubler with parallel stage V_T cancellation on input stages.

becomes negative feedback. The IV curve bends lower from the ideal curve at high I_{load}.

With the decoupling of the gates of input stages from the internal nodes of the main stage, the gate bias is no longer a function of load current. Cross-coupled pair voltage doubler functionality is maintained. The shift from the old IV curve to the new IV curve demonstrates the benefit of the new architecture in resolving self-loading effect from the current load.

3.4.1.2 Clock Phases

In terms of clock phases, there is no new requirement. Since both the gate stage and main stage are cross-coupled pair voltage doubler in nature, nonoverlapping clock phases between CLK1 and CLK2 are the only requirement. Separate buffering technique allows the isolation of noise between the main stage and gate stage, as discussed earlier.

3.4.1.3 Floor Plan, Signal Routing, and Matching

Parallel stage V_T cancellation on the input stage uses two parallel stages with similar structures. These two stages could be easily laid out side by side to allow the best matching of device and, signal flow. All the benefits described in Sec. 3.3.1.4 could be applied here.

3.4.1.4 Power Efficiency

All DC–DC power converters have the start-up or settling requirement. As illustrated in Fig. 3.47, the transition of V_{out} from 0 V to the final target level V_{reg} traverses the IV curve through the low V_{out} region. For the basic cross-coupled pair voltage doubler, the nonideal IV characteristic due to self-loading effects impacts the start-up or recovery speed. With heavy load, there will be a longer settling time for design to be stuck in the lower output current region.

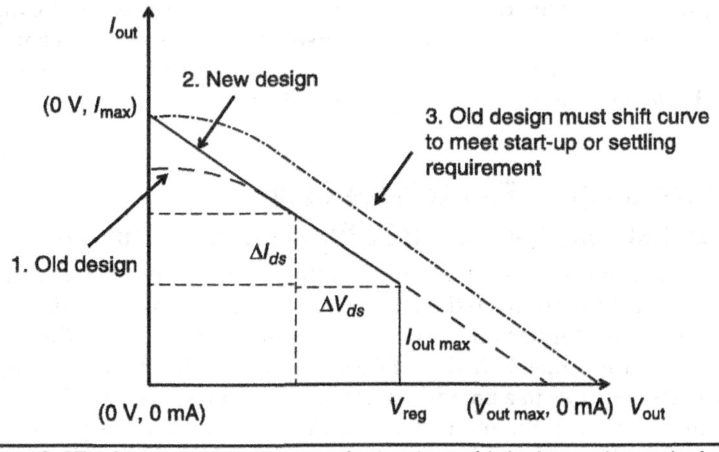

FIGURE 3.47 Start-up or settling requirement on old design and new design.

The newly introduced architecture solves the issues with minimum effort. The *IV* curve of the new design would stay on the ideal slope. No extra power is consumed by the new design. In the old design, the gate loading of N_1 or N_2 is to the main stage coupling capacitance C_1 or C_2. In the new design, the gate loadings are removed from the main stage and are added to the gate stage. The shift of power consumption from main stage to gate stage does not introduce extra power.

Owing to the nonideal *IV* characteristic at low V_{out} in the old design, the charge pump must be upsized to the meet the start-up or settling time specification. As illustrated in Fig. 3.47, curve 1 is the old design using the basic cross-coupled pair voltage doubler. Curve 2 is the new design with parallel stage V_T cancellation on the input stage. Since certain types of charge pumps are designed to drive only capacitive load, the size of the charge pump must be increased to meet the timing specification. Curve 3 is shifted from curve 1 to meet the timing specification.

If parallel stage V_T cancellation were used on the input stages, then the charge pump would not be overdesigned but would still meet both the timing and output current design specifications. Power efficiency benefit has to be evaluated in simulation because of the complexity of nonlinear effects.

3.4.1.5 Area Efficiency

As discussed in Sec. 3.3.1.6 area efficiency can be easily derived from power efficiency. Since both timing and the *IV* curve require transient simulations to understand both circuits and loads, designers can do the trade-off analysis easily to see the benefit of new architecture.

3.4.2 Architecture Limitations

With parallel stage V_T cancellation on input devices of the cross-coupled pair voltage doubler, the nonoverlapping clocks are the only requirement to ensure to cut off reverse charge leakage. In terms of architecture, there is no new concern except for the V_T limitation on the diode-connected output stage transistors.

3.5 Cross-Coupled Pair Voltage Doubler: Parallel Stage V_T Cancellation Optimum Design

In the last two sections, two new architectures were introduced to address the limitations in the cross-coupled pair voltage doubler. One addresses the limitation of V_T of the output stage. The other one addresses the impact of the self-loading effects in the input stage. Both architectures use an additional parallel gate stage, which is similar in structure to the main stage, but try to address different issues. For the optimum single-stage charge pump design, can these two

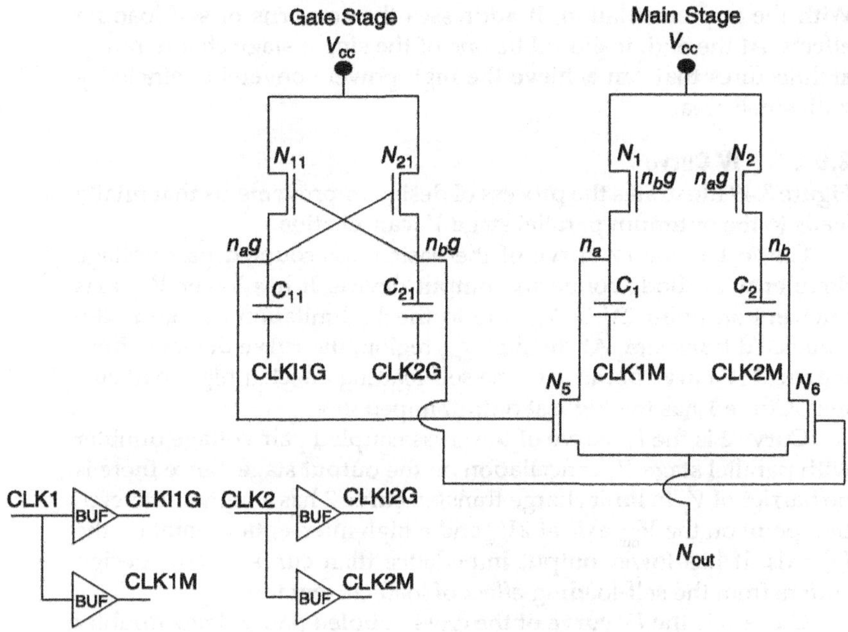

FIGURE 3.48 Cross-coupled pair voltage doubler with optimum parallel stage V_T cancellation.

schemes be combined to achieve the optimum design with the cross-coupled pair voltage doubler structure as a single-stage pump?

Figure 3.48 illustrates the cross-coupled pair voltage doubler with optimum parallel stage V_T cancellation. The main stage delivers power to the output. The gate stage is used for V_T cancellation on both input and output stages. Two stages are similar in structure except for sizing and connection. Loading on the gate stage is usually much smaller than that on the main stage, and it allows smaller sizing in the gate stage. The gates of transistors in the main stages are driven by signals from the gate stage. Node $n_a g$ is connected to the gates of N_2 and N_5, while node $n_b g$ is connected to the gates of N_1 and N_6. The clock drivers for the main stage and gate stage are identical in clock phase. To minimize noise interference between two stages, the clock buffer separation technique is used.

3.5.1 Advantages

In circuit design, simplicity is better. The cross-coupled pair voltage doubler with optimum parallel stage V_T cancellation is simple in structure and clock phase control. It applies V_T cancellation to the input stage to fully receive the input power, and it also applies V_T cancellation to the output stage to fully deliver power to the load. The clocking scheme and clock phase requirements are simple to meet.

With the implementation, it addresses the concerns of self-loading effects. At the end, it should be one of the single-stage charge pump architectures that can achieve the high power conversion efficiency with small area.

3.5.1.1 *IV* Curve

Figure 3.49 illustrates the process of design improvements that finally leads to the optimum parallel stage V_T cancellation.

Curve 1 is the *IV* curve of the basic cross-coupled pair voltage doubler with diode-connected output device. It has lower V_{out} axis interception point $2V_c - V_{tn}$ due to the V_T limitation of the diode-connected transistor. At the high I_{out} region, the curve deviates from the ideal *IV* curve because of the self-loading effect at high load current. Curve 1 has the highest output impedance.

Curve 2 is the *IV* curve of the cross-coupled pair voltage doubler with parallel stage V_T cancellation on the output stage. Since there is no barrier of V_T to limit charge transfer, curve 2 has a higher interception point on the V_{out} axis at $2V_{cc}$ and a high interception point on the I_{out} axis. It has lower output impedance than curve 1. This design suffers from the self-loading effect of load current too.

Curve 3 is the *IV* curve of the cross-coupled pair voltage doubler with parallel stage V_T cancellation scheme applies to both input and

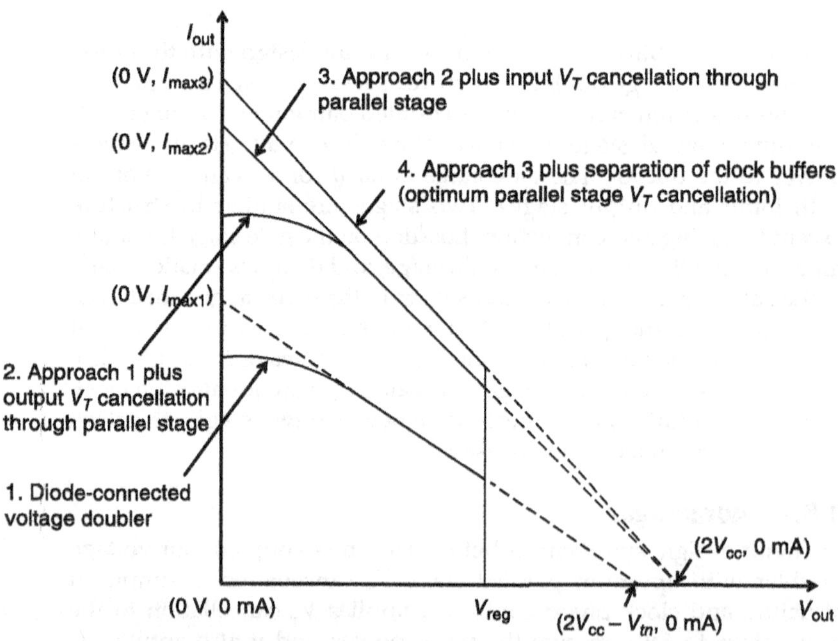

FIGURE 3.49 The *IV* characteristic of the cross-coupled pair voltage doubler with optimum parallel Stage V_T cancellation.

output stages. The self-loading effect due to load is resolved. Curve 3 has constant equivalent output impedance regardless of the low or high output load.

Curve 4 is the *IV* curve of design with optimum parallel stage V_T cancellation. Noise isolation is achieved by the separation of clock buffers between the main stage and gate stage. Compared with curve 3, it has an identical V_{out} axis interception point, but has a higher interception point on the I_{out} axis. Curve 4 has the lowest equivalent output impedance of all.

3.5.1.2 Clock Phase

Only two clocks are required in the design, and with nonoverlapping clock phases. Nonoverlapping clocking phases can be achieved easily by using the design illustrated in Fig. 3.19. Because there is no complicated clock scheme, and clock phases are simple, this design is much simple and more easily achieved over different PVT corners. At the end, simple design usually generates the best result in production.

3.5.1.3 Floor Plan, Signal Routing, and Matching

Figure 3.50 illustrates a floor plan and signal flow for the new design. The gate stage and the main stage are placed side by side. Capacitors are laid out on top while clock buffers are placed on the bottom. Since two stages are similar in nature, routing tracks for both power and signals can be shared and are matched. Signals flow from left to right,

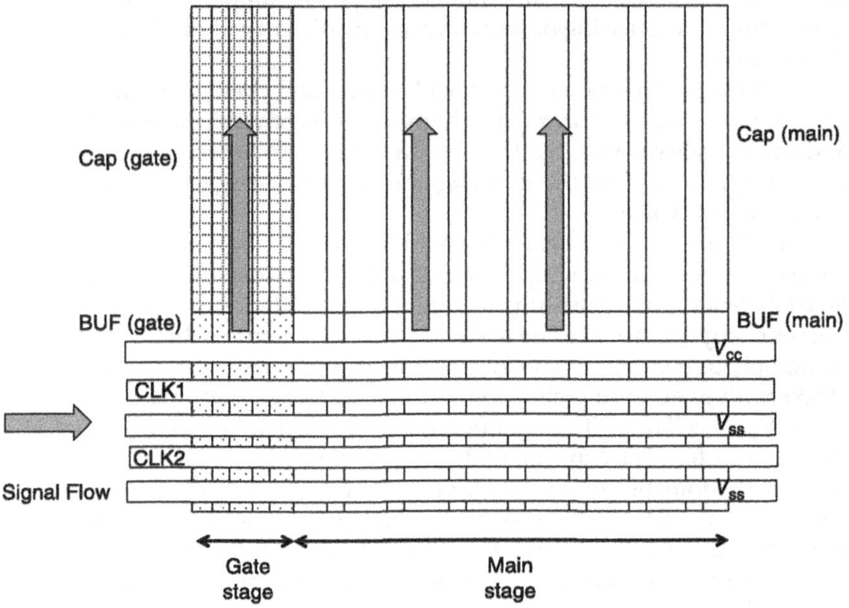

FIGURE 3.50 Floor plan and signal flow of parallel stage V_T cancellation design.

from the gate stage to the main stage. In vertical direction the signals flow from clock buffers to capacitance.

There is inherent delay in signal transitions from the gate stage to the main stage. This delay helps the gate clock to turn on or off early prior to the switching of the internal node of the main stage. It will help to increase the power efficiency of the design.

3.5.1.4 Power Efficiency and Area Efficiency

Figure 3.49 illustrates the *IV* curve of the cross-coupled pair voltage doubler with optimum parallel stage V_T cancellation. For any on-chip switch capacitor power supply design, if there is no inhibition in input and output stages to transfer charge, then the power conversion is already optimized to the maximum. Vice versa, if power conversion efficiency is maximized, to meet the same output power specification less chip area is needed. Area efficiency would be maximized.

3.6 Architecture Summary

In the domain of on-chip DC-DC power converters the Dickson charge pump is the simplest architecture that can elaborate the concept of switch mode power conversion with capacitors. With the scaling of technology, the given chip supplies are scaled down for device reliability and reduction of power consumption. The scaling trend of chip supplies is challenging for charge pump design. Leakage and device reliability concerns would not allow the threshold voltage to be scaled down at the same rate as the power supply. This trend would further reduce the power conversion efficiency of the Dickson charge pump.

With the proliferation of hand held devices and cloud computing, on-chip power conversion requires higher power conversion efficiency and smaller silicon area to allow the overall system to reduce power and design costs. This trend makes the Dickson charge pump less attractive in practice.

Any V_T cancellation scheme would be a better alternative to the Dickson charge pump in improving power conversion efficiency and area efficiency. As discussed in charge pump circuit design, the four-phases charge pump was one of the popular architectures that is used commonly in industries. The four-phase charge pump requires main clocks with overlapping clock phases for precharging and discharging in V_T cancellation. Two additional auxiliary clocks are needed for charge transfer. They are embedded in corresponding main clock phases. The four-phase charge pump is not easy to design. The complexity of multiple clock phases is a challenge across PVT corners. With the additional layout-related issues and power supply integrity, worst case timing margins are required. Under typical conditions, usually there are design margins left unused.

Architectures determine the baseline performance and design complexity. Many complex V_T cancellation architectures have been proposed in the past. As in any other analog designs, simplicity is better.

The cross-coupled pair voltage doubler with optimum parallel stage V_T cancellation presents a very simple V_T cancellation architecture to achieve very high power conversion efficiency. The V_T cancellation has been applied to both input and output stages. The clock scheme is simply nonoverlapping, or clocks could simply cross over at 50% of the transition edge. The clock scheme is simple, and both the gate stage and main stage track with each other over PVT corners. It makes layout floor planning and signal planning simple.

In conclusion, the cross-coupled pair voltage doubler with optimum parallel stage V_T cancellation is a simple pump architecture that could achieve the high power conversion efficiency and less silicon area with given design specification.

3.7 References

1. Dickson, J., "On-chip high-voltage generation NMOS integrated circuits using an improved voltage multiplier technique," *IEEE J. Solid-State Circuits*, vol. SC-11, no. 3, pp. 374–378, June 1976.
2. Pan, F., J. H. Huynh, inventors; SanDisk Corporation, assignee. "Charge Pump with Vt Cancellation Through Parallel Structure." United States patent application US20090302930 A1; 2009 December 10.
3. Pan, F., T. Samaddar. *Charge Pump Circuit Design*. McGraw-Hill Professional, New York, 2006.
4. Kalluru, V. V., S. Yadala, S. Chandrasekarn. Inventors; SanDisk Corporation, assignee. "Circuits for prevention of reverse leakage in vth-cancellation charge pumps." United States patent application US20140077868 A1: 2014 March, 20.

CHAPTER 4

Multistage Charge Pump

In Chap. 3, the design of the single-stage charge pump using the cross-coupled pair voltage doubler is discussed in great detail. In multistage charge pump design, more insights are added to elaborate the key points in charge pump design.

4.1 Positive Charge Pump with Diode Isolation

The positive charge pump with diode isolation is probably one of the most commonly used charge pumps in the field. In Chap. 1, the design of charge pump is analyzed with ideal circuit elements: diodes and capacitors. By replacing diodes with N-channel MOSFET and capacitors with MOSFET capacitors, a practical CMOS positive charge pump is realized.

4.1.1 Design Concept

Figure 4.1 illustrates the basic architecture of the positive charge pump with diode isolation, or the Dickson charge pump.[1] In the figure two single-pump stages are cascaded in series. In each pump stage, there is a diode-connected N-channel MOSFET, such as N_1 or N_2. There is AC coupling capacitance in each pump stage, which is implemented with the actual MOSFET device. The first pump stage receives input from the input power supply V_{cc}. At the output, the diode-connected NMOS isolates the load from the internal stage of the charge pump.

4.1.2 Nonidealities

With real circuit elements, there are nonidealities that would impact the performance, power efficiency, and area efficiency of the design.

4.1.2.1 Parasitic Capacitance

The very first nonideality with the charge pump design is related to the parasitic capacitance associated with critical nodes in each pump stage.

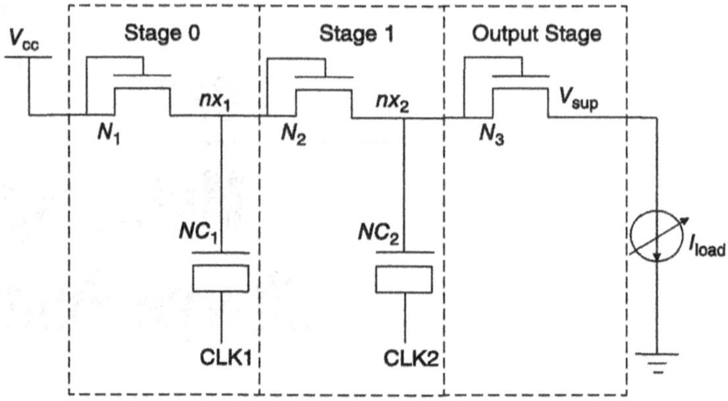

Figure 4.1 Positive charge pump with diode isolation (Dickson charge pump).

The location of the parasitic capacitance has different implications for design.

Figure 4.2 illustrates parasitic capacitance within the pump stage. Two parasitic capacitances are added to the internal nodes of the pump stage: C_{p1} and C_{p2}.

Capacitance C_{p1} is located on the bottom plate of the coupling capacitance NC_1. It is connected between CLK1 and ground. The main components of C_{p1} are the junction and gate S/D overlapping capacitance of the clock driver, interconnect routing capacitance, and parasitic capacitance associated with the bottom plate of device NC_1. The choice of AC coupling capacitance for NC_1 could be either NMOS or PMOS depending on the polarity of the biases. With *NMOS-type* of coupling

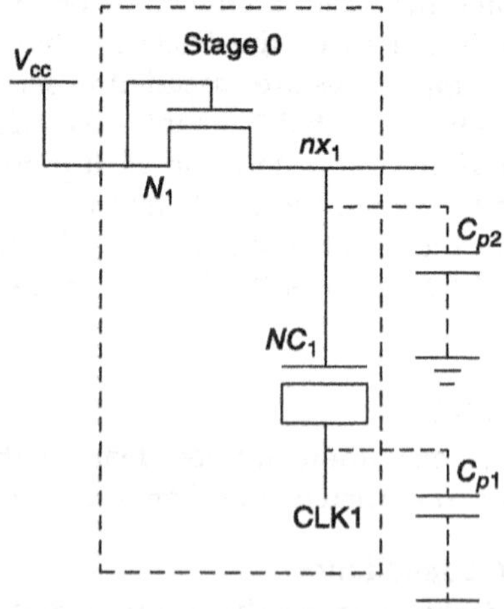

Figure 4.2 Parasitic capacitance within the pump stage.

capacitance, the parasitic capacitance associated with bottom plate of NC_1 would be the depletion capacitance between conducting channel and the $P_{\text{substrate}}$. With the PMOS device as the coupling capacitance, the parasitic capacitance associated with the device would be the depletion capacitance between NWELL and the P substrate.

The addition of C_{p1} to CLK1 has great impact on the design. First, the switching power consumption of the charge pump will increase. This energy is wasted since it is not transferred to the output load. Second, the clock slew rate will be slowed down with the added capacitance. If the duration of slew of CLK1 were relatively small compared with the pump clock period, C_{p1} should not impact the output voltage level or current capability. However, if the duration of slew of CLK1 were slow enough, it would impact the performance of the design. Clock drivers must be optimized to take account of the impact of C_{p1}.

Capacitance C_{p2} is located on the interconnection nx_1 between stages. It is composed of the junction and gate overlap capacitance of diode-connected NMOS in the current pump stage, interconnect capacitance, and junction capacitance and gate capacitance of diode-connected NMOS in the next stage.

The addition of C_{p2} would impact the design performance significantly. Most of the parasitic capacitance associated with C_{p2} is proportional to the device sizes. Capacitance C_{p2} would steal away charge that is supposed to be delivered to the output. Lowering of both the maximum output voltage and the power efficiency is expected. The maximum voltage that could be achieved on stage 0 output nx_1 is given by Eq. (4.1), where k is the coupling ratio (CR).

$$V_{nx1,\max} = V_{cc} - V_T + V_{cc}\frac{C}{C+C_{p2}} = V_{cc}(1+k) - V_T \tag{4.1}$$

$$V_{\max} = V_{cc} - V_T + N(kV_{cc} - V_T) \tag{4.2}$$

The maximum output voltage that could be achieved with N-stage pump design is given by Eq. (4.2). Normally with load current, the maximum voltage that node nx_1 could achieve in the real case is $V_{nx1,\text{real}}$ in equilibrium. The excess charge calculated from Eq. (4.3) could be used to derive the load current shown in Eq. (4.4).

$$Q_{\text{tran}} = C(V_{cc}(1+k) - V_T - V_{n1,\text{real}}) \tag{4.3}$$

$$I_{\text{load}} = \frac{Q_{\text{tran}}}{T_{\text{clk}}} = \frac{C(V_{cc}(1+k) - V_T - V_{nx1,\text{real}})}{T_{\text{clk}}} \tag{4.4}$$

And, if the load current I_{load} is known the output voltage in stage 0 can be found easily, as shown in Eq. (4.5).

$$V_{nx1}(I_{\text{load}}) = V_{cc}(k+1) - V_T - \frac{I_{\text{load}}T_{\text{clk}}}{C} \tag{4.5}$$

As the number of pump stages N increases, the maximum voltage with load current at output for the N-stage positive charge pump could be predicted by

$$V_{max} = (V_{cc} - V_T) + N\left(kV_{cc} - V_T - \frac{I_{load}T_{clk}}{C}\right) \tag{4.6}$$

4.1.2.2 Threshold Voltage and Overdrive Voltage

The second nonideality also is related to the threshold voltage V_T and overdrive voltage V_{ov} of the diode-connected NMOS in each pump stage. The voltage V_T of NMOS will increase as the internal node voltage increases along the pump stages. The overdrive voltage V_{ov} is a function of the load current. Since the effect of V_T has been included in the preceding calculation, the maximum output voltage for an N-stage charge pump is revised in Eq. (4.7) with V_{ov} included. As the number of pump stages N increases, the impact on V_{max} from of V_T and V_{ov} will be multiplied by N. As I_{load} increases, its impact on V_{max} from V_{ov} and the switching capacitance capability will be multiplied as N increases.

$$V_{max} = V_{cc} - V_{tn} - V_{ov} + N\left(kV_{cc} - V_T - V_{ov} - \frac{I_{load}T_{clk}}{C}\right) \tag{4.7}$$

4.1.2.3 Parasitic Resistance

The third nonideality is related to the parasitic resistance on power supplies, the internal interconnect between stages, the ESR of capacitance, and other routing resistance associated with drivers that can cause *IR* drop or delay in operations. Parasitic resistances cause the maximum voltage achievable in Eq. (4.7) to be lower than the target. The effective maximum clock frequency achievable would be lower. Hand modeling of physical layout or back-annotation should be included in simulations to check the impact and improvement for design.

4.1.3 *IV* Curve Analysis

Figure 4.3 illustrates the *IV* characteristic of the multistage positive charge pump. As the number of pump stages N increases, the *IV* curve rotates counterclockwise. From the plot, the intercept point between *IV* on the V_{out} axis gives the maximum output voltage V_{max} achievable. As the number of pump stages N increases, V_{max} will shift higher proportionally, as shown in Eq. (4.8). The intercept point between *IV* and the I_{out} axis gives the maximum achievable output current I_{max}, which is given in Eq. (4.9). For a single-stage pump, the predicted current is given in Eq. (4.4). As the number of pump stages N increases, it will take the charge N clock cycles from the input

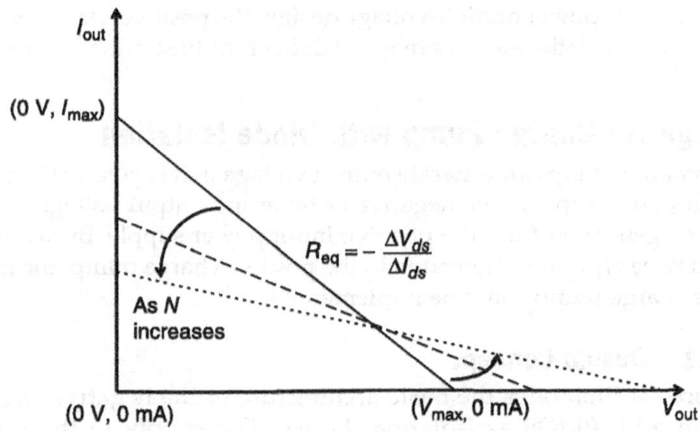

FIGURE 4.3 *IV* of *N*-stage positive charge pump.

supply to reach the load. With loss in the path, the real output current will be less than anticipated.

$$V_{max} = V_{cc} - V_T + N(kV_{cc} - V_T) \tag{4.8}$$

$$I_{max} = \frac{C(V_{cc}(1+Nk) - (N+1)(V_T + V_{ov})}{NT_{clk}} \tag{4.9}$$

$$R_{out} \approx N\left(\frac{T_{clk}}{C} + \frac{1}{g_m}\right) \tag{4.10}$$

The slope of the *IV* curve is inversely proportional to the output impedance of design R_{out}, which is given in Eq. (4.10) as function of the number of pump stages N, clock period T_{clk}, coupling capacitance C, and g_m of the diode-connected element. As the number of stages N increases, R_{out} increases proportionally. Equations (4.8) to (4.10) identify all the key parameters that impact multistage positive charge pump design.

4.1.4 Conclusion

The multistage positive charge pump (or Dickson charge pump) with diode-connected isolation devices is simple to implement, and it has fewer design risks. As highlighted in Eqs. (4.8) to (4.10), compared with the single-stage pump design, the factors that are multiplied by the number of pump stages N are V_T, V_{ov}, and g_m. With the diode acting as both the charge transfer and isolation device, not only does the device V_T limit the charge transfer, but also it has lower g_m (transconductance) to transfer charge. Large dimension is required for coupling capacitance and transconductance of the device. Power efficiency would be degraded by threshold voltage V_T and larger parasitic capacitance associated with larger devices.

For high-power or high-voltage design, the positive charge pump with diode isolation devices may not deliver the best value for design.

4.2 Negative Charge Pump with Diode Isolation

The charge pump can generate output voltage that is greater than the input power supply. The negative or inverting output voltage could also be generated from the positive input power supply. By swamping device type and connection in the positive charge pump, the negative charge pump could be implemented.

4.2.1 Design Concept

Figure 4.4 illustrates the basic architecture of the negative charge pump with PMOS as isolation diodes. The supply to the input stage is Gnd. Normal NMOS could not be used in the negative pump stages since the S/D junction of NMOS would be forward biased with negative voltage on either the source or the drain node. In Fig. 4.4, there are two single pump stages cascaded in series. In each pump stage, the isolation device is implemented with PMOS, such as P_1–P_3. The configuration of diode-connected PMOS allows positive charge to be dumped to ground, and negative charge could remain and propagate to the load. The NWELL connection of each PMOS is tied to the output of the preceding stage, which has higher potential than the current stage. Also NWELL could be tied to V_{cc} or ground for simplicity. Since the S/D junction of PMOS with respect to NWELL is reverse biased, it is safe for negative voltage operation. AC coupling capacitance could be implemented with any MOSFET capacitor with the gate of each device tied to the internal node.

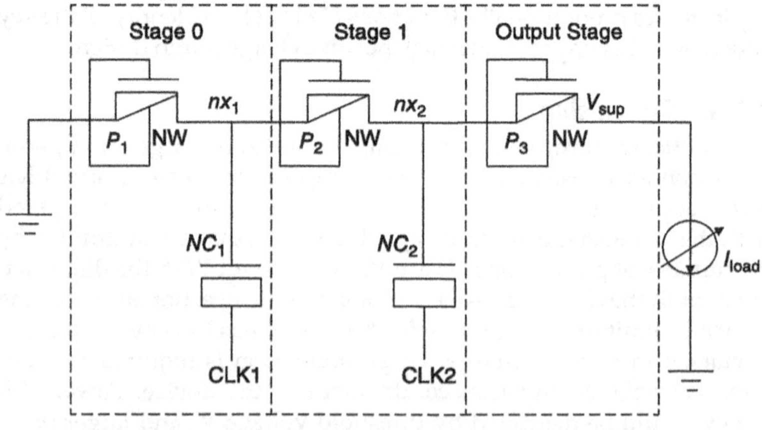

Figure 4.4 Negative charge pump with PMOS as isolation diodes.

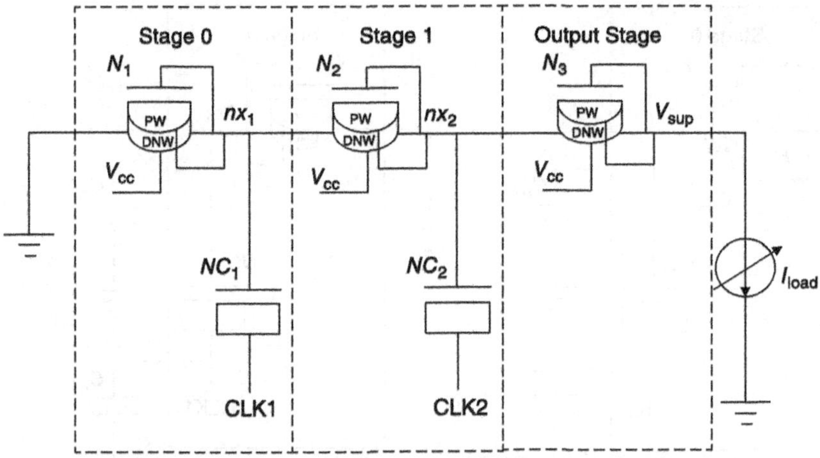

FIGURE 4.5 Negative charge pump with triple WELL NMOS isolation diode.

Figure 4.5 illustrates the basic architecture of the negative charge pump with triple WELL NMOS acting as the isolation diode. In the illustration, there are two single pump stages are cascaded in series, plus an output stage. In each pump stage, the ideal diode is implemented with the actual NMOS sitting in the triple WELL, such as N_1–N_3. The configuration of NMOS allows positive charge to be dumped to ground, and negative charge can propagate to the output. The P-WELL is tied to the output of the current pump stage since it is the most negative potential in current stage to prevent body bias. Deep N-WELL DNW is connected to V_{cc} or ground to isolate the P-WELL.

4.2.2 Nonidealities

With real circuit elements, there are a few nonidealities related to the parasitic capacitance, threshold voltage, and parasitic resistance in design.

4.2.2.1 Parasitic Capacitance

Similar to the positive charge pump with isolation diode, the size and location of parasitic capacitance have a significant impact on performance in many aspects.

Figure 4.6 illustrates the parasitic capacitance located in the negative charge pump with either PMOS diode or triple WELL NMOS diode configuration.

The impacts of parasitic capacitances C_{p1} and C_{p2} on the negative charge pump are similar to their impact on the positive charge pump, as highlighted in Sec. 4.1.2.1. The most negative voltage that could be achieved with a one-stage negative charge pump with

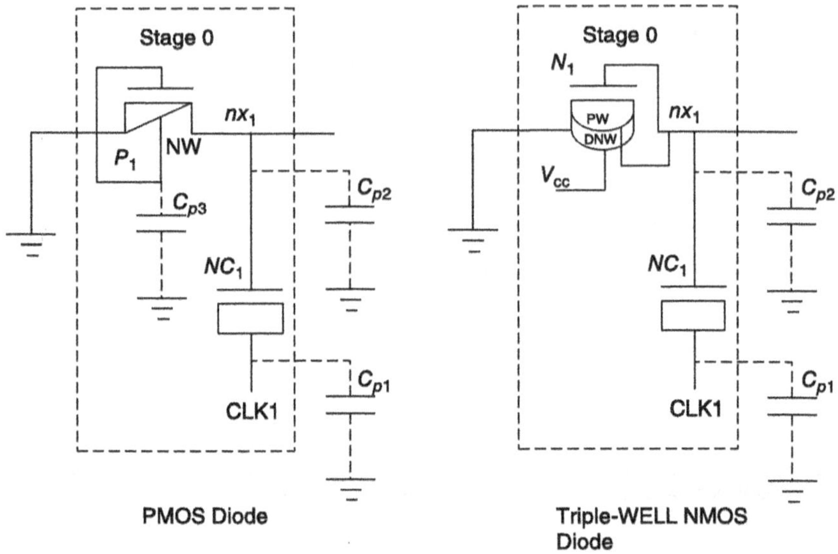

FIGURE 4.6 Parasitic capacitance in negative charge pump.

diode isolation is given by Eq. (4.11), where k is defined as the CR between C and C_p.

$$V_{nx1,\,\min} = |V_t| - V_{cc}\left(\frac{C}{C+C_p}\right) = |V_T| - kV_{cc} \tag{4.11}$$

$$V_{\min} = -N(kV_{cc} - |V_T|) \tag{4.12}$$

The minimum output voltage that could be achieved by N-stage design is given by Eq. (4.12). With load current, both node NX_1 and NX_2 can only achieve $V_{nx1,\,\mathrm{real}}$ in equilibrium; then any excess charge given in Eq. (4.13) would be used to deliver the load current within clock period T_{clk}, as shown in Eq. (4.14).

$$Q_{\mathrm{tran}} = C(|V_T| - kV_{cc} - V_{nx1,\,\mathrm{real}}) \tag{4.13}$$

$$I_{\mathrm{load}} = \frac{Q_{\mathrm{tran}}}{T_{\mathrm{clk}}} = \frac{C(|V_T| - kV_{cc} - V_{nx1,\,\mathrm{real}})}{T_{\mathrm{clk}}} \tag{4.14}$$

And if the load current I_{load} is known, the minimum output voltage at the output of a single-stage negative pump could be easily derived by Eq. (4.15);

$$V_{nx1}(I_{\mathrm{load}}) = |V_T| - \left(kV_{cc} + \frac{I_{\mathrm{load}}T_{\mathrm{clk}}}{C}\right) \tag{4.15}$$

As the number of stages increases for the negative charge pump, the minimum voltage at output with load current for the N-stage negative charge pump could be predicted from Eq. (4.16).

$$V_{\min} = N\left(|V_T| - kV_{cc} - \frac{I_{\text{load}}T_{\text{clk}}}{C}\right) \tag{4.16}$$

4.2.2.2 Threshold Voltage and Overdrive Voltage

The second nonideality is related to threshold voltage V_T and overdrive voltage V_{ov} of the diode-connected PMOS or NMOS in each stage. Compared with the positive charge pump, with either PMOS or triple WELL NMOS, since the body potential of passing transistor can be tied to local bias, V_T would not suffer from increasing body effect down the stages in the negative charge pump design. Since the effect of V_T has been included in the preceding calculation, the minimum output voltage was revised in Eq. (4.17) to include V_{ov}. As the number of pump stages increases, the effect of V_T, V_{ov}, and the switch resistance would be multiplied.

$$V_{\min} = N\left(-kV_{cc} + |V_T| + V_{ov} - \frac{I_{\text{load}}T_{\text{clk}}}{C}\right) \tag{4.17}$$

4.2.2.3 Parasitic Resistance

Parasitic resistance impacts the negative charge pump performance in a similar way as it impacts the positive charge pump.

4.2.3 *IV* Curve Analysis

Figure 4.7 illustrates the *IV* curve of N-stage negative charge pump. Since it is for negative potential, the absolute value of the voltage is taken. As the number of negative pump stages N increases, the *IV*

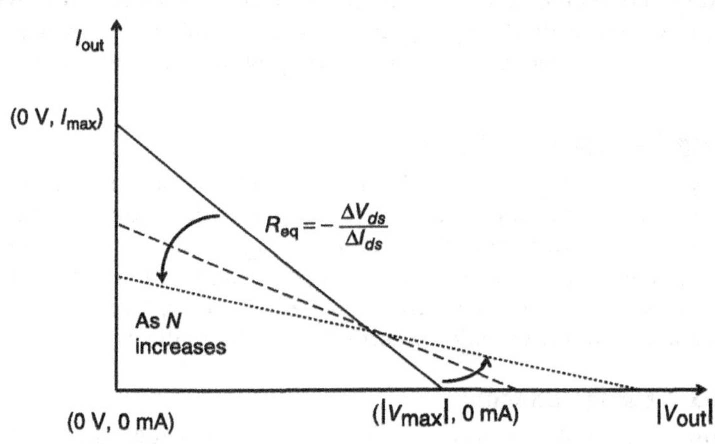

Figure 4.7 *IV* of negative N-stage charge pump.

curve rotates counterclockwise. From the plot, the intercept point between *IV* and the V_{out} axis determines the most negative output voltage V_{max} achievable. As the number of pump stages *N* increases, $|V_{max}|$ shifts higher proportionally, as given by Eq. (4.18). The intercept point between the *IV* curve and I_{out} axis gives the maximum achievable output current I_{max}, which is given in Eq. (4.19). Similar to the *N*-stage positive charge pump, as the number of pump stages *N* increases, it takes *N* clock cycles to move charge across the *N*-stage negative charge pump. With loss in the path, real output current would be less than anticipated.

$$|V_{max}| = N(|V_T| - kV_{cc}) \tag{4.18}$$

$$I_{load\ max} = \frac{NC[-C/(C+C_p)V_{cc} + |V_T| + V_{ov}]}{NT_{clk}} \tag{4.19}$$

$$R_{out} = N\left(\frac{T_{clk}}{C} + \frac{1}{g_m}\right) \tag{4.20}$$

The output impedance R_{out} given by Eq. (4.20) is represented by the inverse of the slope of the *IV* curve. As the number of negative pump stages increases, the output impedance of the charge pump will be increased proportionally.

As *N* increases in Fig. 4.7, the *IV* curve rotates counterclockwise, as illustrated.

4.2.4 Conclusion

The negative charge pump with diode isolation is relatively easy to implement and has fewer design risks. The negative charge pump does have the advantage over the positive charge pump of having new devices that do no have to suffer from increasing body bias. As highlighted in Eqs. (4.18) to (4.20), the critical parameters that impact the performance are V_T, V_{ov}, T_{clk}, and C. As the number of pump stages increases, the impact in single-stage pump would be multiplied.

4.3 Parallel Stage V_T Cancellation

Many parallel stage V_T cancellation architectures[2] are introduced in Chap. 3 for single-stage pump design. The concept could be easily extended to the implementation of multistage positive charge pump or negative charge pump design to achieve higher power efficiency and higher area efficiency. All the benefits of the parallel stage V_T cancellation concept would be carried over automatically.

4.3.1 Design Concept

Figure 4.8 illustrates the basic implementation of the multistage positive pump with parallel stage V_T cancellation. In the figure, there are two

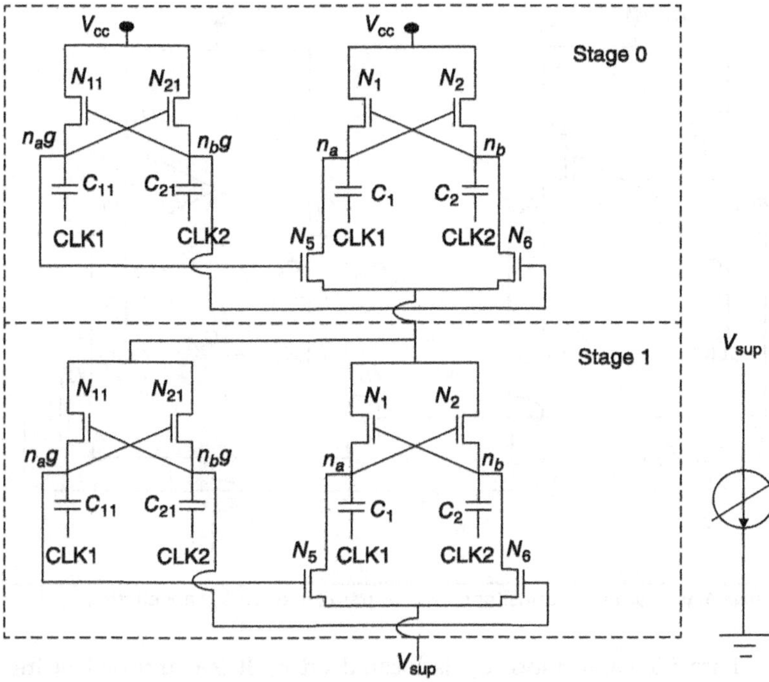

FIGURE 4.8 Multistage positive charge pump with parallel stage V_T cancellation.

single-stage pumps cascaded in series. The output of stage 0 becomes the input supply for stage 1. Each pump stage is a cross-coupled pair voltage doubler with parallel stage V_T cancellation on the output stage. In this particular implementation, each stage operates independently of the other except for the input/output supplies between the two stages.

4.3.2 Nonidealities

With real design elements, a few nonidealities show up in the design.

4.3.2.1 Parasitic Capacitance

With parallel stage V_T cancellation, the parasitic capacitances are illustrated in Fig. 4.9 for the left half of circuits.

Parasitic capacitance C_{p1} is located on the output of the clock driver and the bottom plate of the coupling capacitance. It comprises the junction capacitance and gate S/D overlapping capacitance of clock drivers, and the interconnect capacitance and parasitic capacitance associated with the bottom plate of the device capacitance C_1. Choice of device type for C_1 determines the size of associated parasitic capacitance. This is similar to other architectures. The impact of C_{p1} on pump design is similar to what is discussed in Sec. 4.1.2.1.

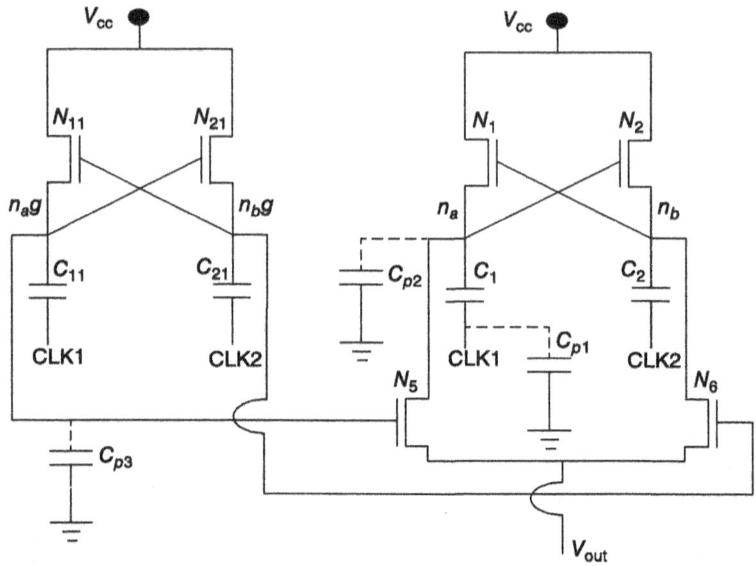

FIGURE 4.9 Parasitic capacitance within parallel stage V_T cancellation.

Parasitic capacitance C_{p2} is located on n_a. It is composed of the junction and gate overlap capacitance of output stage N_5, interconnect capacitance, junction and gate/source overlap capacitance N_1, and gate capacitance of N_2. Parasitic capacitance C_{p2} would lower the coupling efficiency and reduce the output voltage and current drivability as highlighted in Sec. 4.1.2.1.

Parasitic capacitance C_{p3} is located on the output of gate stage $n_a g$. Parasitic capacitance C_{p3} would lower the CR from CLK1 to node $n_a g$. Lowering the maximum value of $n_a g$ would reduce the charge transfer efficiency of the main stage through N_5.

$$V_{na\,\max} = V_{cc}\left(1+\frac{C}{C+C_p}\right) = V_{cc}(1+k) \tag{4.21}$$

$$V_{\max} = V_{cc} + N(kV_{cc}) \tag{4.22}$$

The maximum voltage that could be achieved at the output of stage 0 is given by Eq. (4.21), where k is defined as the CR. The maximum output voltage that could be achieved with N-stage pump design is predicted by Eq. (4.22). With the load current, node n_a in stage 0 could only achieve $V_{na\,\text{real}}$ in steady state. The excess charge in Eq. (4.23) is used to supply to the load defined in Eq. (4.24).

$$Q_{\text{tran}} = C(V_{cc}(1+k) - V_{na\,\text{real}}) \tag{4.23}$$

$$I_{\text{load}} = \frac{Q_{\text{tran}}}{T_{\text{clk}}} = \frac{C[(1+k)V_{cc} - V_{na\,\text{real}}]}{T_{\text{clk}}} \tag{4.24}$$

Vice versa if the load current I_{load} is known, it is easy to identify the maximum output voltage at the output of stage 0 with Eq. (4.25):

$$V_{na}(I_{load}) = V_{cc} + \left(kV_{cc} - \frac{I_{load}T_{clk}}{C}\right) \quad (4.25)$$

As the number of pump stages increases, the maximum voltage with load current for the N-stage positive charge pump could be predicted as

$$V_{max} = V_{cc} + N\left(kV_{cc} - \frac{I_{load}T_{clk}}{C}\right) \quad (4.26)$$

4.3.2.2 Threshold Voltage and Overdrive Voltage

With parallel stage V_T cancellation, the V_T of passing NMOS would be canceled in each pump stage. With the gate voltage controlled by a parallel stage, V_{ov} would not be a limiting factor in charge transfer on the main path. The maximum output voltage is given in Eq. (4.26) with the parallel stage V_T cancellation scheme.

4.3.2.3 Parasitic Resistance

The discussion related to parasitic resistance from Sec. 4.1.2.3 could apply to the N-stage positive charge pump with parallel stage V_T cancellation. Each gate stage output needs to drive the pass gate in the main stages. The parasitic R going to N_1 or N_2 may not be as critical as the impact from the parasitic C due to routing.

4.3.3 *IV* Curve Analysis

Figure 4.10 illustrates the *IV* characteristic of N-stage parallel stage V_T cancellation on the output stage. The design follows the similar trend to Fig. 4.3 for the N-stage positive charge pump.

$$V_{max} = V_{cc} + N(kV_{cc}) \quad (4.27)$$

$$I_{max} = \frac{C[V_{cc} + N(kV_{cc})]}{NT_{clk}} \quad (4.28)$$

$$R_{out} = N\left(\frac{T_{clk}}{C} + \frac{1}{g_m}\right) \quad (4.29)$$

Equation (4.27) gives the maximum output voltage V_{max} that could be achieved. Without the impact of V_T and V_{ov}, Eq. (4.27) delivers the maximum output voltage achievable if compared with Eq. (4.8), particularly with the multiplication factor N in the formula. Equation (4.28) defines the maximum output current I_{max} that could be delivered. The benefit of V_T cancellation is clear if compared with Eq. (4.9). Equation (4.29) defines the output impedance of the charge pump. In this design, g_m of the passing element is

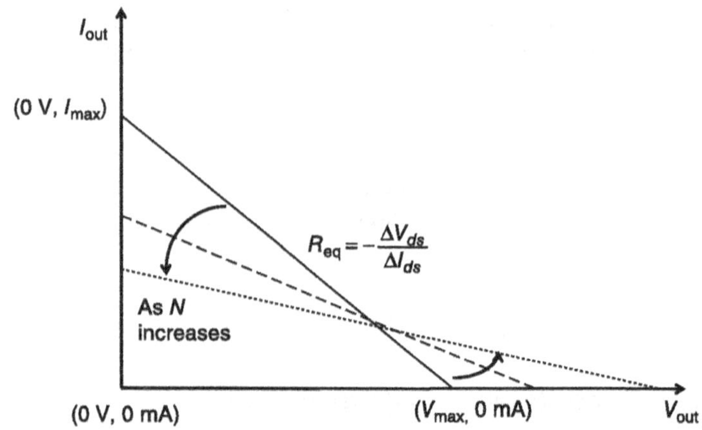

Figure 4.10 The *IV* curve of *N*-stage parallel stage V_T cancellation on the output stage.

maximized with parallel stage design. With better charge transfer, T_{clk} could be scaled down accordingly. And R_{out} would be improved with respect to Eq. (4.10).

Compared to the early design without V_T cancellation, all the metrics from Eqs. (4.8) to (4.10) show significant improvement.

4.3.4 Conclusion

Compared to the performance metrics of multistage the positive charge pump with isolation diode (or Dickson charge pump) given by Eqs. (4.8) to (4.10), multistage parallel stage V_T cancellation shows improved performance as the projection. The design is reasonably simple and has little risk in design and in silicon.

4.4 Self-adaptive Charge Pump

The self-adaptive charge pump[3] is a special multi-stage charge pump utilizing the parallel stage V_T cancellation scheme. The architecture would automatically reconfigure itself to the proper number of stages optimized for target regulation level.

4.4.1 Design Concept

Figure 4.11 Illustrates the basic architecture of the *N*-stage self-adaptive charge pump. Each pump stage is configured similar to the configuration of the parallel stage V_T cancellation design. There are differences in connection. Instead of getting the output from the preceding main stage output as the input supplies for the main and gate stages, the gate stage receives its input supply from the preceding gate stage's output. All gate stages are connected in a chain, and all the main pump stages

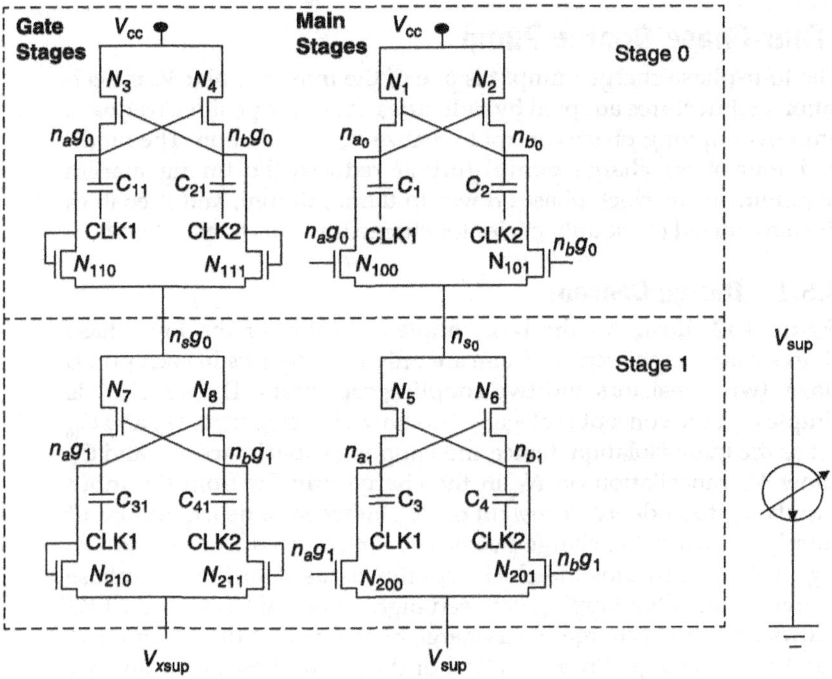

FIGURE 4.11 The *N*-stage self-adaptive charge pump.

are connected in a chain. The gain stages have fixed load in operation. The main stages are self-adapted to the optimum number stages needed to maintain the target regulation level. Last few stages in the main stages could be completely bypassed if the regulation level is in the lower range. The gate stages are shown as diode-connected, or they could have additional V_T cancellation if needed. Voltage V_{sup} is the target output supply for the design. Voltage V_{xsup} is the output of gate stage. It should track the V_{sup} regulation level directly or indirectly.

There are several advantages to the self-adaptive charge pump. If the output regulation varied over a large range, the self-adaptive charge pump would automatically settle to the optimum number of pump stages required at any given time. The last few stages of the main stage are simply short and bypassed if regulation level is low. This leads to the minimum charge pump output impedance seen at any given condition. The output impedance of the charge pump is self-adaptive to the variation in output target level. There is much less output noise because bypassed stages in the main stage chain act as the filter capacitance. The minimum charge p output impedance would lead to the most optimal charge pump performance at any given time. Analysis of the self-adaptive charge pump would be a good exercise for interested readers.

4.5 Four-Phase Charge Pump

The four-phase charge pump[4–6] is one of the most popular V_T cancellation architectures adopted by industries. It adopts positive feedback with overlapping phase control to realize V_T cancellation. The modified four-phase charge pump[7] further reduces the timing margin requirement for clock phases down to the minimum, and it enables the maximized clock duty cycles for charge transfer in operation.[4]

4.5.1 Design Concept

Figure 4.12 illustrates the basic implementation of the four-phase charge pump architecture. There are only four devices in each pump stage: two transistors and two coupling capacitors. The structure is simple and the concept is elegant. As shown in stage zero, N_{10} and C_{10} act as the main isolation device and coupling capacitance; N_{11} and C_{11} allow V_T cancellation on N_{10} in the charge transfer from the input supply V_{cc} to node nx_1. They cut off N_{10} between node nx_1 and input supply V_{cc} when the charge needs to be transferred from nx_1 to nx_2. Figure 4.13 illustrates the clock waveforms used in the four-phase charge pump. Overlapping between high phases of CLK1 and CLK2 allows proper precharge or discharge of the gate of the passing element in each stage. Precise control of the clock phases over all PVT corners is challenging. To cover the worst-case PVT corners, the normal PVT corner could have too much margin left unused.

The detailed analysis of four-phases charge pump design is illustrated in Charge Pump Circuit Design.[8]

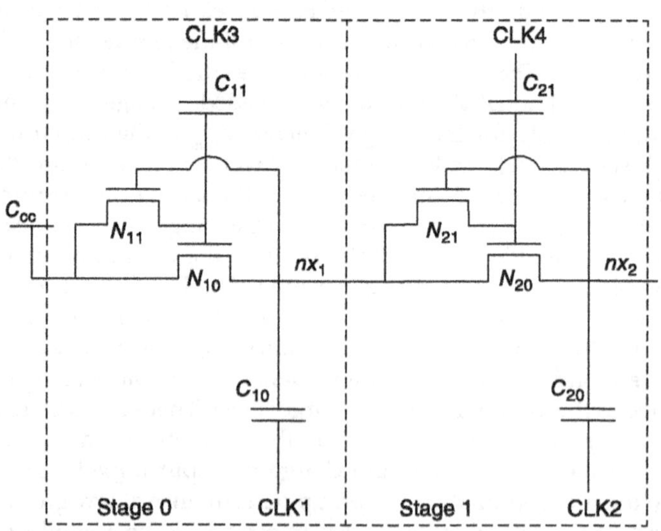

FIGURE 4.12 Four-phase charge pump.

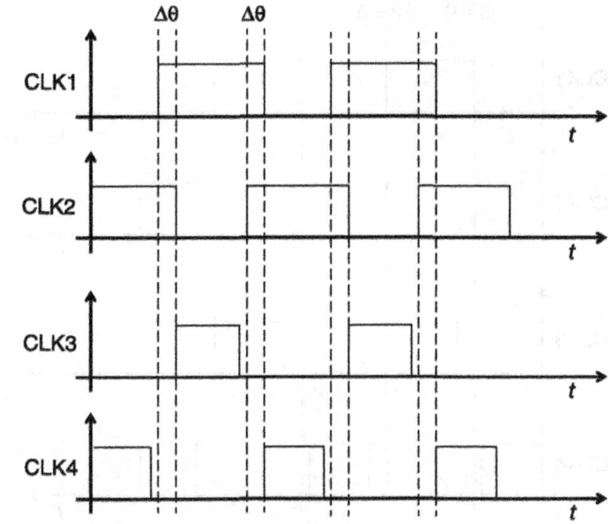

Figure 4.13 Clock waveforms of the four-phase charge pump.

4.5.2 Four-Phase Charge Pump without Clock Phase Overlap

Figure 4.14 illustrates a modified four-phase charge pump design without phase overlap.[4] As shown in Fig. 4.15, there is no need to overlap CLK1 and CLK2. The added diode in each pump stage guarantees that precharge will occur before the clocking of CLK3 or CLK4. Discharge of the gate of the pass element is less of a concern. The phase requirements on CLK3 and CLK4 are simple.

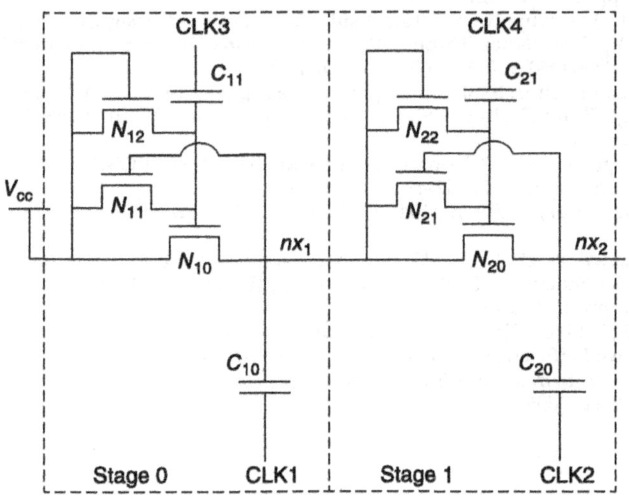

Figure 4.14 Four-phase charge pump without clock phase overlap.

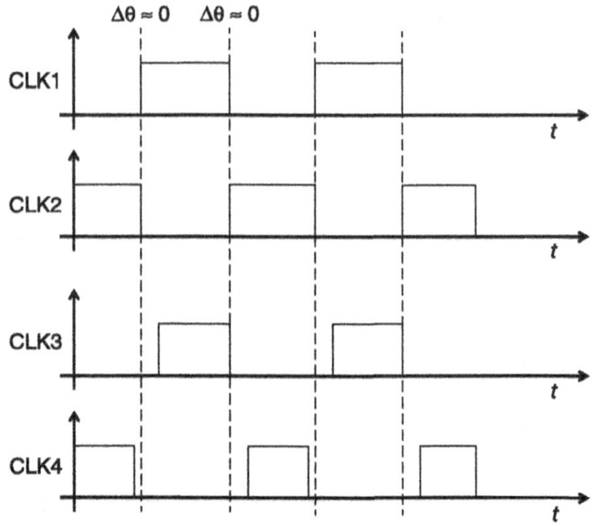

Figure 4.15 Clock waveforms of the modified four-phase charge pump.

With relaxed conditions on the phase control requirement, the modified four-phase charge pump without clock phase overlap can run at a higher clock frequency and have higher efficiency.

4.6 References

1. Dickson, J. "On-chip high-voltage generation NMOS integrated circuits using an improved voltage multiplier technique," *IEEE J. Solid-State Circuits*, vol. SC-11, no. 3, pp. 374–378, June 1976.
2. Pan, F., J.H. Huynh, inventors; SanDisk Corporation, assignee. Charge Pump with V_T Cancellation Through Parallel Structure. United States patent application US20090302930 A1; 2009 December 10.
3. Pan, F., inventor; SanDisk Corporation, assignee. "Self-Adaptive Multi-Stage Charge Pump." United States patent application US20100019832 A1; 2010 January 28.
4. D'Arrigo, S. et al., "A 5V-only 256k Bit CMOS Flash EEPROM", ISSCC Digest of Technical Papers, pp. 132–133, Feb., 1989.
5. Dham, V. et al., "A 5V-only E2PROM using 1.5µ lithography", ISSCC, pp. 166–167, 1983.
6. Umezawa, A. et al., "A 5V-Only Operation 0.6um Flash EEPROM with Row Decoder Scheme in Triple-Well Structure", IEEE J. Solid-State Circuits, vol. 27, no. 11, pp. 1540–1546, Nov. 1992.
7. Pan, F., T. Pham. "Four phase charge pump operable without phase overlap with improved efficiency." U.S. patent no. 7,030,683, April 2006.
8. Pan, F., T. Samaddar. *Charge Pump Circuit Design*. McGraw-Hill Professional, New York, 2006.

CHAPTER 5

Charge Pump Clock Driver

5.1 Pump Clock Driver Design

The pump clock driver design is an integral part of the charge pump design. On one hand, the pump clock amplitude and pump clock slew rate impact the performance of the charge pump tremendously. On the other hand, the charge pump output noise and supply variation are linked directly to the design of the pump clock driver.

5.1.1 Supply of Pump Clock Driver

The pump clock driver is intended to drive large capacitance. It has to source or sink a large amount of current over a short time. The power supplies of the pump clock driver have to be strong to minimize the impacts to the settling time constant of pump clocks. Without a doubt, the input power supplies, such as V_{cc} and GnD, would have the least amount of impedance on the chip. They can source or sink a large amount of current without degradation of the supply voltages.

In designing the charge pump, block placement and power bus planning are two of the critical aspects of the charge pump design. The block placement is composed of global and local placements. The global placement determines the width of global power buses required from power pads to charge the pump block. Block placement also determines the width of output supplies from the charge pump to the load across the chip. The local placement impacts the planning of local power networks, signal flow, and routings.

In terms of global placement of the charge pump design, it is critical to know the design specification. What are the input power supply range and input power capability, and what is the charge pump output load current profile of the design?

Figure 5.1 illustrates the circuit diagram of the charge pump design with the input power network included. The dotted square delineates the charge pump circuit design. The components outside the dotted square are related to the routing resistances of power

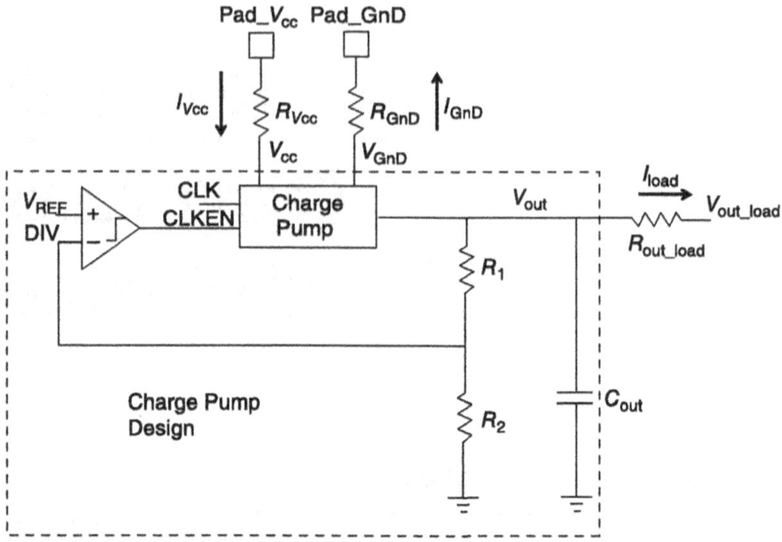

FIGURE 5.1 Charge pump design with input power networks.

supplies and the output routing resistance. Input power supplies at power pads PAD_V_{cc} and PAD_GnD are at ideal levels. With the associated power network impedances R_{vcc} and R_{GnD}, supply currents I_{Vcc} and I_{GnD} would cause internal V_{Vcc} and V_{Gnd} to deviate from the ideal supplies. Normally I_{Vcc} and I_{GnD} are much larger than I_{load}, and charge pump performance is sensitive to its input supplies. Special care must be taken in planning the power network with the proper estimates of I_{Vcc} and I_{GnD} consumed by the charge pump and other circuits in operation.

$$I_{Vcc} >> I_{load} \quad \text{and} \quad I_{GnD} >> I_{load} \tag{5.1}$$

Equation (5.1) illustrates a basic relationship among I_{Vcc}, I_{GnD} and I_{load}. Regardless of the charge pump design and configuration, the current efficiency would always be less than 1. As the number of pump stages increases, the current efficiency decreases. The value of I_{Vcc} or I_{GnD} of the charge pump would be much greater than I_{load} in a normal design. The impact of the *IR* drop on V_{Vcc} supply is highlighted in Eq. (5.2) because of the large amplitude of I_{Vcc}. Similarly, the *IR* drop would apply to the ground supply due to large ground current.

$$V_{cc} = V_{PD\_V_{cc}} - I_{Vcc}R_{Vcc} \tag{5.2}$$

It would be intuitive to place the charge pump close to the power pad to minimize the input supply *IR* drop with minimum chip size impact.

Similarly, $R_{out, load}$ and I_{load} should be analyzed in the planning of the charge pump design. With the priority to place the charge pump near the input supply pads, $R_{out, load}$ would be increased because of the longer routing distance. With the maximum load current, there would be a voltage difference between the point of load and the output of the charge pump:

$$V_{diff} = V_{out} - V_{out\_load} = I_{load} R_{out\_load} \tag{5.3}$$

The difference is described in Eq. (5.3). If design could not tolerate the error, either widening the output bus or moving the feedback sampling point near load would minimize this error in regulation.

5.1.2 Delay of Internal Buffering Stage

In designing the charge pump clock driver, the delay in internal pump clock buffering stages should be small enough not to impact the duty cycle of the pump clock and delay of buffer followed.

$$I_{ds} = \frac{1}{2}\mu_n C_{ox} \frac{W}{L}(V_{cc} - V_T)^2 \tag{5.4}$$

The saturation region dominates the major portion of the signal transition time of the pump clock buffer. To use NMOS as an example, Eq. (5.4) gives I_{ds} of NMOS in high to low transition on the output of the clock buffer. With the gate of NMOS fully biased at V_{cc}, I_{ds} has a second-order relationship with input power supply V_{cc}. For output switches from low to high, PMOS follows similar second-order relationship with input power supply V_{cc}, and Eq. (5.4) could be modified with the associated PMOS parameters.

$$C_{load} = C_{ox} W_{N+1} L_{N+1} + C_{ov} W_N + C_{junction} W_N S_{junction} + C_{jsw}(W_N + 2S_{junction}) \tag{5.5}$$

Equation (5.5) illustrates the capacitive load of the clock buffer during switching. The total capacitive load seen by the internal clock buffer at stage N includes the input capacitance of the next stage $(N+1)$ and all the junction capacitance and overlapping capacitance of the clock buffer at stage N.

$$t_d = \frac{C_{load}(0.5V_{cc})}{I_{ds}} \propto \frac{L^2}{V_{cc}} \tag{5.6}$$

Equation (5.6) illustrates the derivation of the transition delay between the internal buffering stages of the clock drivers. Assuming the switching threshold voltage of buffer is at one-half of the input supply voltage V_{cc}, and I_{ds} of the current stage operates in saturation region, then the stage delay t_d is proportional to the square of channel length L, and it is also inversely proportional to the input supply voltage V_{cc}.

Since the intermediate buffering stage delay t_d should be as small as possible, the minimum channel length L_{min} of the given technology should be chosen. With L_{min}, it allows a smaller width of transistor to be used and consumes less silicon area for design. On the other hand, maximizing the input supply voltage to the charge pump clock drivers would reduce the stage delay, and it can also reduce the width of the transistors in the pump clock buffer design and overall silicon area.

In general, minimizing the intermediate pump clock buffer stage delay allows better timing margins for the charge pump clock design over PVT corners. It demands less silicon area for pump clock buffers and minimizes unwanted power consumption due to crowbar current during signal transitions.

5.1.3 Final Stage of Clock Buffer and Buffering Technique

The performance of the charge pump is a strong function of the charge pump clock. Both the amplitude and the settling behavior of the charge pump clock will impact the final performance of the charge pump design. The amplitude of the charge pump clock in transient is a function of the slew rate and linear settling time constant of the final buffered pump clock.

5.1.3.1 Slew Rate

Figure 5.2 illustrates the waveforms of the charge pump clock V_{clk} in transition, where V_{clk} is the buffered clock signal from V_{in} and V_{clk} is designed to drive an equivalent load capacitance C_{eq}. The clock period is t_{cycle}. Voltage V_{clk} must reach its final target level through transition in half of a clock period. If input clock V_{in} toggles between 0 V and V_{cc}

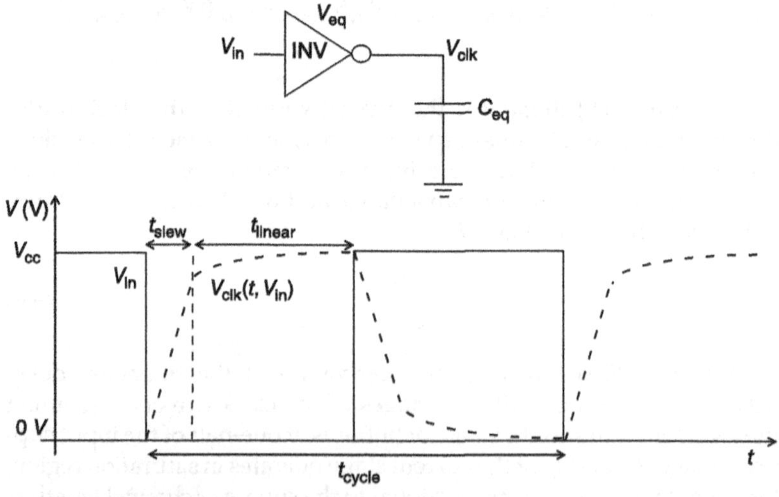

FIGURE 5.2 Waveforms of charge pump clock.

with the transitions being approximated as step functions, then the transient response of V_{clk} per transition cycle could be divided into two operating regions—the slew region and linear settling region, as illustrated in Fig. 5.2.

$$t_{slew} = \frac{C_{eq} V_{slew}}{I_{ds}} \tag{5.7}$$

In the slew region, t_{slew} is the switching time for the final stage buffer before it enters the linear settling region. The maximum pull-up or pull-down I_{ds} of the final clock buffer determines the slew rate of transition. Equation (5.7) gives the quantitative figure of t_{slew} as a function of C_{eq}, V_{slew}, and I_{ds} of design. Voltage V_{slew} is the swing of output voltage before either the pull-up or pull-down transistor in the final buffer enters the linear region; I_{ds} limits the maximum t_{slew} that could be achieved.

With a large output load to drive, logic effort for charge pump clock buffering would help to reduce the overall transition delay. Figure 5.3*a* illustrates an example of logic effort for the charge pump clock buffering. It is assumed that the load capacitance C_{eq} is driven by an inverter. The input capacitance of the inverter is normalized to 1, and the capacitive load is normalized to be k^4 ($k > 1$). The propagation delay of the inverter with fan-out of 1 is assumed to be t_d, and

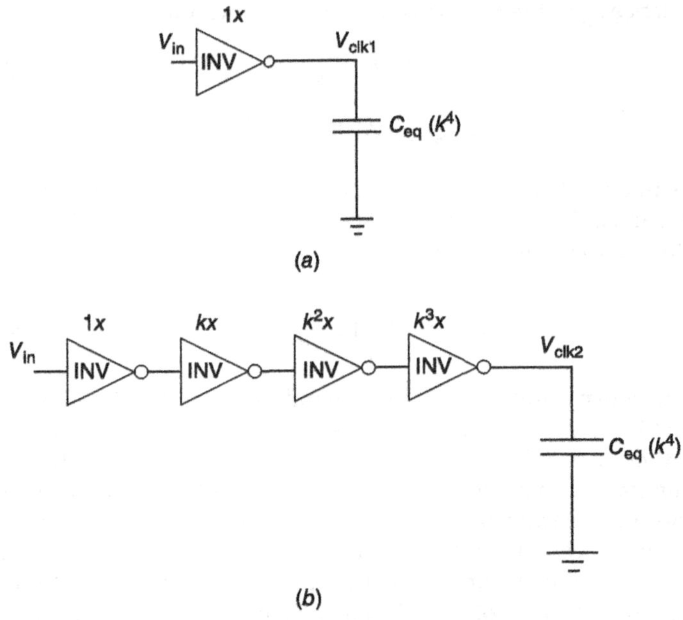

FIGURE 5.3 Logic effort for charge pump clock buffering.

the delay for case (*a*) could be calculated by Eq. (5.8). It is obvious the final delay would grow at 4th order with *k*.

$$t_{\text{slew1}} = k^4 t_d \tag{5.8}$$

Logic effort for pump clock buffering is illustrated in Fig. 5.3*b*. With the addition of three extra buffers in the size indicated, the final delay change to a linear relationship with *k*, as shown in Eq. (5.9); t_{buffer} is the intrinsic delay through the buffer due to its own loading. The total propagation delay between input and output is given by Eq. (5.10). The choice of *k*, which is related to the final choice of the number of buffering stages, should be optimized by the delay, duty cycle, power consumption, and area requirement of the pump clock buffer design.

$$t_{\text{delay2}} = kt_d \tag{5.9}$$

$$t_p = 4kt_d + 3t_{\text{buffer}} \tag{5.10}$$

5.1.3.2 Linear Settling

Once transistors in the final buffering stage enter the triode region, transition of V_{clk} will enter the linear settling region, as indicated in Fig. 5.2. Linear settling is dependent on the time constant, which is given by Eq. (5.11), where C_{eq} is the total equivalent capacitance seen by the final pump clock buffer and R_{eq} is the linear region output impedance r_{ds} of the transistor in the final clock buffer.

$$\tau = R_{\text{eq}} C_{\text{eq}} \tag{5.11}$$

$$\Delta V_{\text{clk}}(t) = \Delta V(1 - e^{-t/\tau}) \tag{5.12}$$

Within the linear settling region, ΔV_{out} is a function of *t*, time constant τ, and ΔV_{clk}, as shown in Eq. (5.12), where ΔV is the voltage swing and ΔV_{clk} is the change in the output clock in the linear settling region.

$$t_{\text{linear}} = -\tau \ln\left(1 - \frac{\Delta V_{\text{clk}}}{\Delta V}\right) \tag{5.13}$$

For linear settling behavior, it would be practical to allow some error in the output voltage V_{clk} to have some meaningful timing constraint. Equation (5.13) is derived from Eq. (5.12) to show the linear settling time with respect to the output error allowed. Because of the logarithmic relationship, the delay time t_{linear} exhibits asymptotic behavior, as seen in Fig. 5.4. The table in Fig. 5.4 shows a normalized settling time over τ with respect to the percentage of target voltage change that output voltage V_{clk} has reached. The graph on the right illustrates the asymptotic behavior of linear settling.

$\Delta V_{out}/\Delta V$	t/τ
50.00%	0.6931472
60.00%	0.9162907
70.00%	1.2039728
80.00%	1.6094379
90.00%	2.3025851
91.00%	2.4079456
92.00%	2.5257286
93.00%	2.65926
94.00%	2.8134107
95.00%	2.9957323
96.00%	3.2188758
97.00%	3.5065579
98.00%	3.912023
99.00%	4.6051702

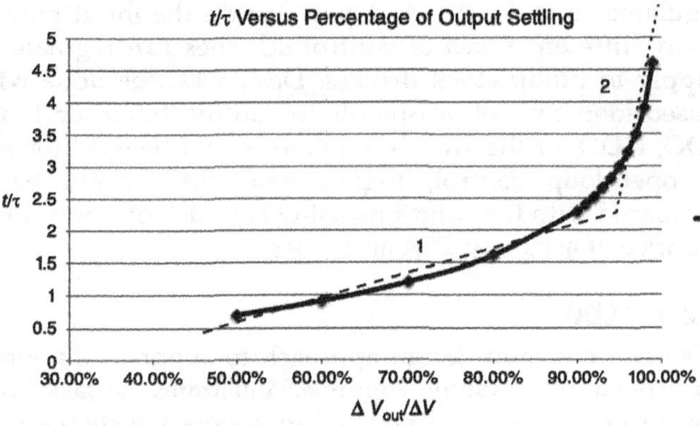

FIGURE 5.4 Linear settling time of charge pump clock.

Before 85% of the final target change ΔV is reached, the normalized delay could be approximated by a dotted line 1 with shallow slope. For settling within the final 15% of the target change ΔV, the linear settling relationship could be approximated by dotted line 2 with a much steeper slope. In this region, to achieve an additional 1% gain in amplitude would require a significant amount of time and effort. The tradeoff between pump clock amplitude and pump clock frequency must be analyzed to maximize the gain of the design.

5.2 Feedback Control

When one is optimizing the charge pump design, one of the design challenges is the suppression of the output noise and PSR with respect to input supply variation. With a given chip design, the input power supply is normally defined to have ±10% variation from the nominal value the chip resides in. As a consequence, both the charge pump average DC output level and output noise would vary from the input supply variation.

There are two ways to solve the issues. Either the design can allocate large decoupling capacitance to lower the output noise, or it can suppress the pump clock amplitude variation to minimize the impact from input supply variation.

Large decoupling capacitance on the output of the charge pump can filter the noise. If the silicon area were not a concern, it would be a viable approach. In many situations, the silicon area is limited and devoting a large silicon area to decoupling capacitance is cost-prohibitive.

The other solution is to address the design of the supply to charge pump clock buffer. If the amplitude of charge pump clock could be less dependent on the input power supply variation, the performance of the overall charge pump system would have better PSR and noise. The supply of the charge pump clock should be regulated to make it less dependent on the input power supply. Many different types of control schemes can regulate the input supply to pump clock drivers. Design can be done with either-closed loop control or open-loop control. In closed-loop control, LDO, HDO, or the switch regulator could be used for regulation. In open-loop control, higher bandwidth could be achieved compared with the solutions using closed-loop control with slight sacrifice of accuracy, PSR and noise.

5.2.1 LDO

LDO is a common design approach to suppress the input supply noise or supply variation. Figure 5.5 illustrates the basic implementation of LDO that could be used to supply the charge pump clock driver. With proper design, LDO can suppress input supply variation up to target bandwidth.

As highlighted in Chap. 1, power efficiency and stability concerns make LDO design challenging.

$$P_{loss} = (V_{cc} - V_{sup})(I_{load} + I_{fb}) \tag{5.14}$$

Equation (5.14) highlights the power loss across PMOS transistor P_1. Normally, V_{sup} is set to be the lower bound of the input supply V_{cc} range. If the input supply were at a typical level or near the upper bound, power loss would be large with LDO. Similarly for power

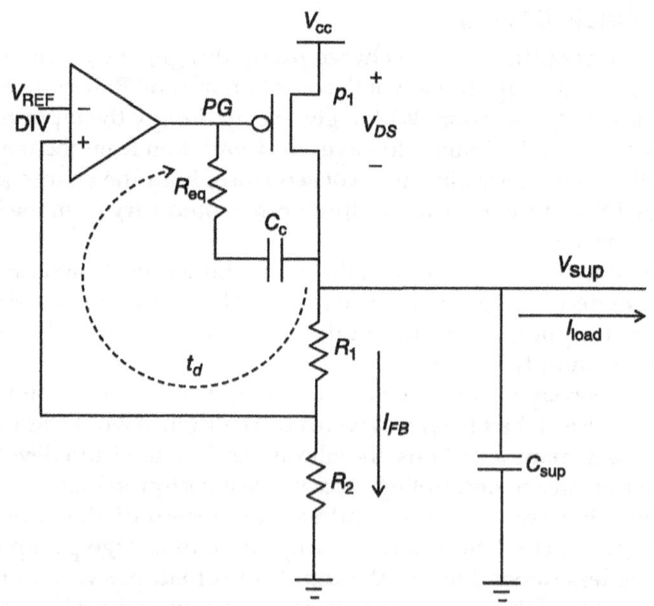

FIGURE 5.5 LDO for pump clock driver supply.

efficiency given in Eq. (5.15), it is optimum if V_{cc} is close to V_{sup}. Power efficiency would drop if V_{cc} shifted away from its lower bound.

$$\text{Power efficiency} = \frac{V_{sup}(I_{load} + I_{FB})}{V_{cc}(I_{load} + I_{FB}) + P_{quiescent}} \tag{5.15}$$

Stability analysis and compensation for LDO are discussed in many textbooks. The details are not covered here. For stability, the compensation could be applied either externally or internally. Compensation on output of the regulator would require large decoupling capacitance and large power consumption in error amplifier. It would have good PSRR. Compensation internally on the output of the error amplifier is a challenge for the load dump response and PSRR.

5.2.2 HDO

HDO is another very popular power regulator that can be used to suppress the input supply noise or minimize supply variation. Figure 5.6 illustrates the basic implementation of HDO.

As discussed in Chap. 1, compared with LDO, HDO has lower power efficiency. This is particularly worse under low-load or no-load conditions. The headroom of the error amplifier determines the minimum power supply voltage required to operate HDO.

$$V_{cc_{min}} = V_{sup} + V_{ov1} \tag{5.16}$$

$$V_{high,\,min} = V_{sup} + V_T + V_{ov1} + V_{ov2} \tag{5.17}$$

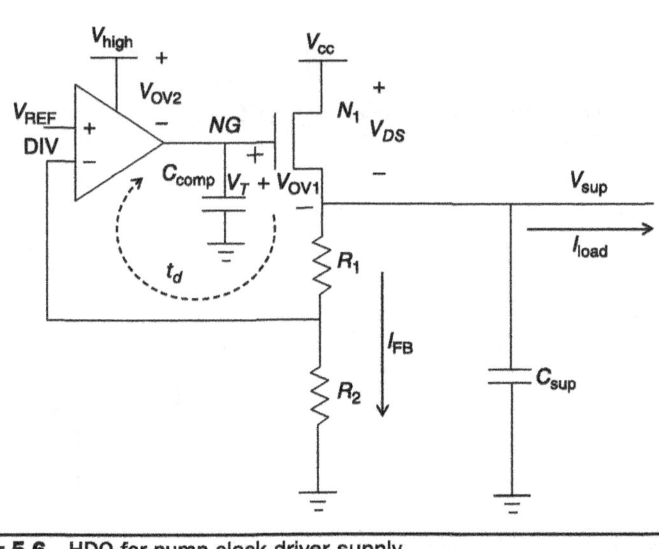

Figure 5.6 HDO for pump clock driver supply.

Equation (5.16) determines the minimum V_{cc} to keep N_1 in the saturation region. Equation (5.17) sets the minimum V_{high} needed to maintain the open-loop gain. Separate power generation is needed for V_{high} if Eq. (5.17) exceeds V_{cc}. If depletion NMOS with negative V_T were available, it would be easy to meet Eq. (5.17) with V_{cc}.

As described in Chap. 1, except for lower power efficiency and lower open-loop gain, HDO has many advantages. It is easy to design and easy to compensate. The whole system can be treated as a single-pole system. In terms of the load dump response, HDO offers better response. Not only is the bandwidth of HDO large, but also I_{ds} of N_1 increases automatically if load dump occurs.

5.2.3 Switch Regulator

With closed-loop designs, both LDO and HDO would require compensation for stability. With compensation, lowering of bandwidth is expected if one is comparing to uncompensated design. Slower response time determines large decoupling capacitance. The concept of the switch regulator would be ideal to achieve the highest bandwidth, which would minimize the silicon area.

Figure 5.7 illustrates an implementation of switch regulator design for the pump clock driver supply. The switch regulator is made of a high-speed comparator, digital buffers, passing element, and feedback network. Except for the comparator and feedback network, all circuits operate in the digital domain to fully utilize the highest bandwidth available under the given technology.

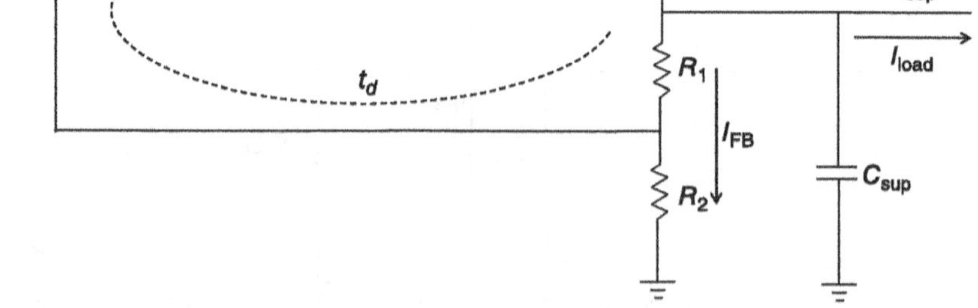

FIGURE 5.7 Switch regulator for pump clock driver supply.

There is no compensation in switch regulator design. Because of the nature of switching, the output will have switching noise around target level.

$$T_{\text{off}} = \frac{\int_0^{t_d} (I_{ds} - I_{\text{load}})\, \partial t}{I_{\text{load}}} + t_d \tag{5.18}$$

$$T_{\text{on}} = \frac{I_{\text{load}} t_d}{I_{ds} - I_{\text{load}}} + t_d \tag{5.19}$$

$$f_{\text{sw}} = \frac{1}{T_{\text{on}} + T_{\text{off}}} \tag{5.20}$$

Equations (5.18) and (5.19) give T_{off} and T_{on} of the switch regulator. Both are a function of I_{load}, I_{ds} and t_d. The switch frequency of the switch regulator is given by Eq. (5.20). As I_{load} approaches 0, the switching frequency f_{sw} drops. As I_{load} approaches I_{ds}, the switching frequency f_{sw} drops too. Switch regulation is based upon a combination of PWM and PFM.

Figure 5.8 illustrates the operations of the switch regulator under load current I_{load}. With low load current I_1, t_{duty1} is short to satisfy the load within T_{period1}. With higher load current I_2, t_{duty2} would be increased to pass more charge to satisfy higher load I_2 within T_{period2}. Over the course of the transitions, both the clock duty and clock period vary.

With high bandwidth of digital circuits and no compensation needed in the switch regulator loop, the loop delay t_d could be a few orders lower than in any LDO designs under same current budgets. Smaller t_d will reduce the decoupling requirement for the load dump.

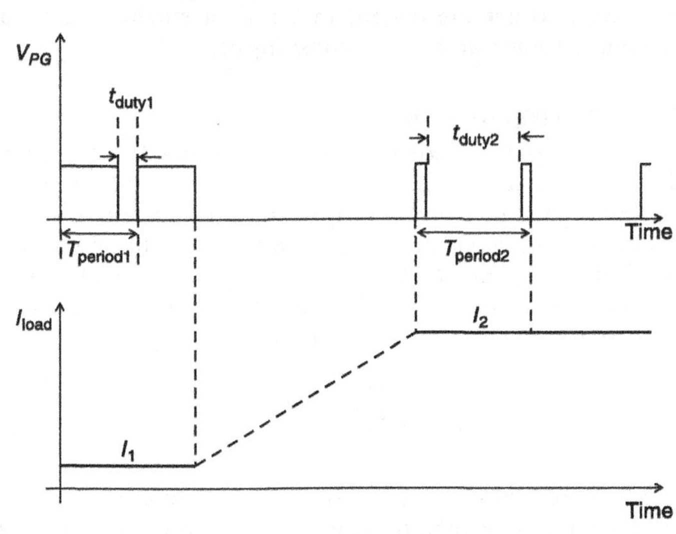

FIGURE 5.8 Regulation of the switch regulator.

In terms of power consumption, the switch regulator has two components: quiescent power from the comparator and feedback network and switching power from digital switching. Minimizing the capacitive loading to the comparator would help to reduce quiescent power through the comparator. The switching power P_{sw} is related to switching frequency f_{sw}, input supply voltage V_{cc}, and total capacitance C, as shown in Eq. (5.21).

$$P_{sw} = CV_{cc}^2 f_{sw} \tag{5.21}$$

$$\Delta V_{output} = \frac{\int_0^{td} (I_{ds} - I_{load}) \partial t}{C_{sup}} + \frac{I_{load} t_d}{C_{sup}} \tag{5.22}$$

To reduce switching power consumption, either C or f_{sw} should be reduced. Smaller C means the reduction of buffer sizing. It leads to increased loop delay t_d. With larger loop delay t_d, Eq. (5.22) illustrates that the output noise ΔV_{output} would increase proportionally. Slower f_{sw} would lead to larger t_d and larger ΔV_{output} based on Eqs. (5.18) to (5.22).

Equation (5.22) illustrates the output noise of the switch regulator. With fixed load current, the output of the switch regulator will switch at fixed frequency f_{sw} with output noise equal to ΔV_{output}. For LDO or HDO, fixed load current the output voltage will be stable. However, in real applications, seldom is the load current fixed in operation. With dynamic load current, the output voltage is expected to be noisy. Digital circuits in design are expected to tolerate power supply noise.

The switch regulator could be combined with other types of regulators, to form either digital assisted analog design or analog assisted digital design[1] to get the benefit in terms of smaller silicon area, reduced output noise, and power consumption.

5.2.4 Open-Loop Control

As long as a negative feedback scheme is utilized to regulate the supply to the charge pump clock drivers, a certain amount of decoupling capacitance on the output supply is required. The minimum decoupling capacitance C_{sup} required by the given design can be derived by Eq. (5.23), where ΔI_{load} is the instantaneous change of load current, t_d is the loop delay, and ΔV_{noise} is the output noise acceptable by design specification.

$$C_{sup} = \frac{\int_0^{td} \Delta I_{load} \, \partial t}{\Delta V_{noise}} \tag{5.23}$$

If the charge pump clock driver could tolerate large supply noise as the long as the total transferred charge did not suffer within the given cycle, then open-loop control could be used to further reduce the decoupling capacitance requirement compared with the closed-loop design.

Figure 5.9 illustrates open-loop control of the power supply for the charge pump clock driver. The bulk of the design is very similar to HDO design. The major difference is that the output is connected to the source of NMOS transistor N_2. Both N_1 and N_2 share the same gate and drain biases.

In Fig. 5.9, V_{INT} is in the negative feedback control loop, and its voltage level is defined by Eq. (5.24). The dimension of N_2 is N times a multiple of the dimension of N_1, and N_2 is configured as the source follower.

$$V_{INT} = V_{REF}\left(1+\frac{R_1}{R_2}\right) \tag{5.24}$$

$$V_{sup} \approx V_{INT} \qquad (\text{if } I_{load} = NI_{bias}) \tag{5.25}$$

Owing to current mirroring, V_{sup} would be equal to V_{INT} if Eq. (5.25) is held true. With the variation of I_{load}, voltage V_{sup} would fluctuate near V_{INT} based on the IV curve characteristic of N_2. For the charge pump clock driver design, the load current is predictable in operation, and the variation of V_{sup} is minimized. For the biasing of N_2, it carries the same benefit as the original HDO to be able to respond instantaneously with load dump to minimize the size of the decoupling capacitance. On the other hand, since the decoupling capacitance is not in the loop of regulation, it simplified the compensation of V_{INT} and the overall design.

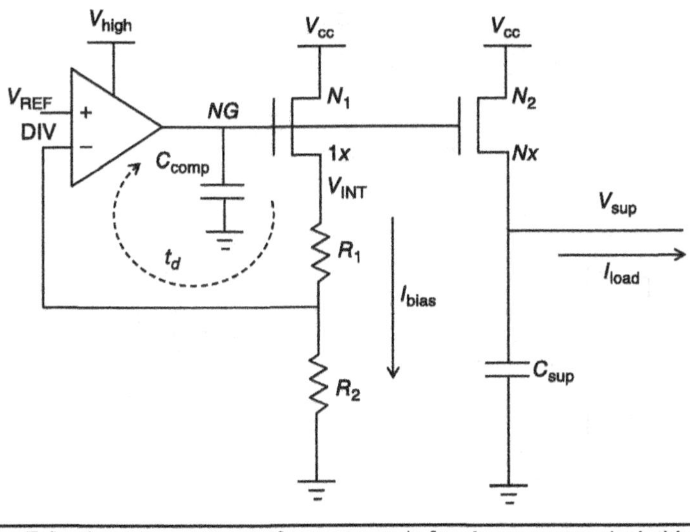

Figure 5.9 Open-loop control of power supply for charge pump clock driver.

5.3 Pump Clock Driver with Stacking Capacitors

Higher chip power supply would increase the power efficiency of the charge pump design. The generation of pump clock amplitude greater than the chip power supply in practice has both area and power conversion efficiency benefits.

Figure 5.10 illustrates an implementation of the pump driver with clock amplitude doubling. The core of the design is the cross-coupled pair voltage doubler in the center. In steady state, Node *NA* switches between V_{cc} and $2V_{cc}$, and node *NB* switches between V_{cc} and $2V_{cc}$ in opposite phase. To generate $2V_{cc}$ amplitude clock, node *NA* and node *NB* are supplied to inverters driven by the clock with the opposite clock phase.

Figure 5.11 illustrates the input and output clock waveforms. Input clock waveforms are CLK1 and CLK2. They switch between 0 and V_{cc} with nonoverlapping clock phases. Output clock waveforms are CLK02 and CLK20. Ideally they would switch between 0 and $2V_{cc}$ without loading. With capacitive load, the maximum output clock amplitude would be lower than $2V_{cc}$.

Figure 5.12 illustrates an implementation of cross-coupled pair voltage doubler using pump clocks with doubled clock amplitude. The outputs from pump clock doubler are fed directly into the clock inputs of cross-coupled voltage doubler.

5.3.1 Charging Technique

Normally if the pump clock driver drove a capacitor *C*, the amount of charge it could supply with a clock amplitude of V_{cc} is given by

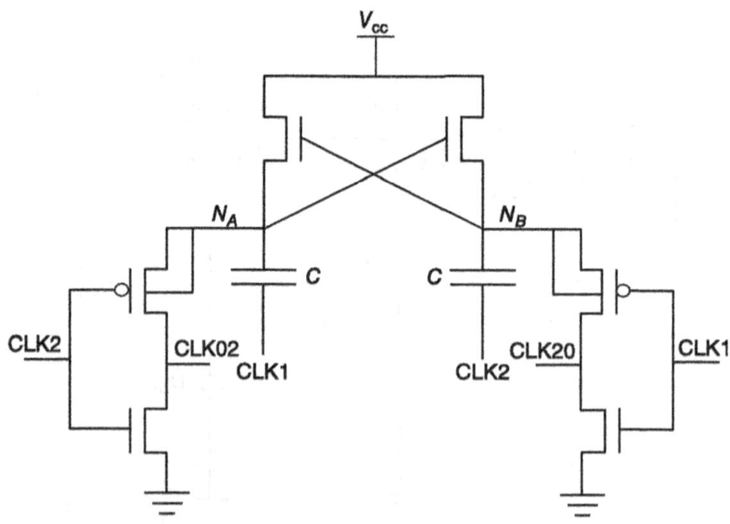

Figure 5.10 Pump clock driver with clock amplitude doubling.

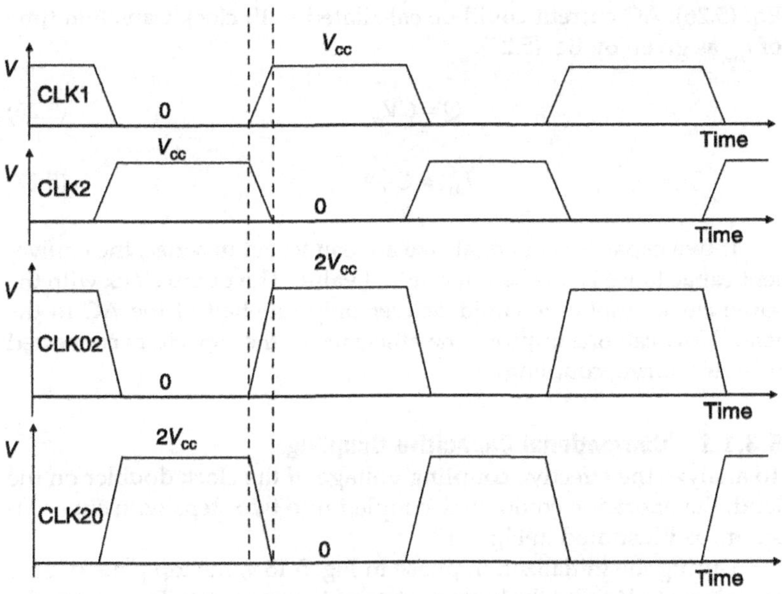

FIGURE 5.11 Clock waves from pump clock amplitude doubler.

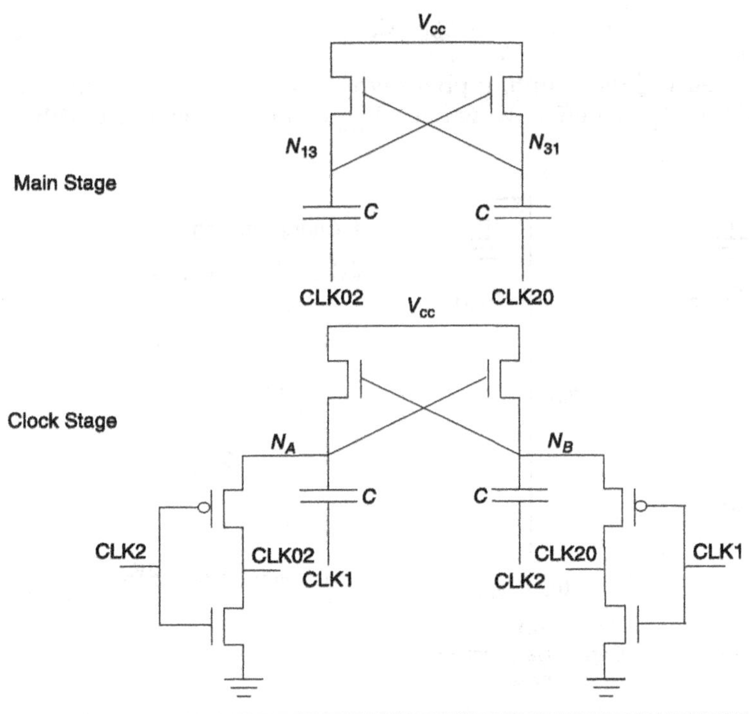

FIGURE 5.12 Voltage doubler with $2V_{cc}$ clock driver.

Eq. (5.26). AC current could be calculated with clock transition time of t_{sw} as given by Eq. (5.27).

$$Q = CV_{cc} \tag{5.26}$$

$$I_{AC} = C\frac{V_{cc}}{t_{sw}} \tag{5.27}$$

If two capacitors of equal size are connected in series, the equivalent capacitance is one half of original value. The pump clock with the same clock amplitude could deliver only one-half of the AC to the load. How can one improve the efficiency of the capacitors connected in series during coupling?

5.3.1.1 Conventional Capacitive Coupling

To analyze the effective coupling voltage of the clock doubler on the load, the operation could be decoupled into two steps with superposition, as illustrated in Fig. 5.13.

During the initialization phase in Fig. 5.13 *a*, the top plate of C_1 is initialized to V_{cc}, and the bottom plate of C_1 is set to 0 V. To simplify the analysis, both plates of C_{load} are set to 0 V. The total amount of charge stored on C_1 is given by Eq. (5.28). The charge stored in C_{load} is 0.

$$Q_{c1} = C_1 V_{cc} \tag{5.28}$$

During the coupling phase when the clock is switching from 0 V to V_{cc}, the effective voltage V_{mid} could be calculated with the

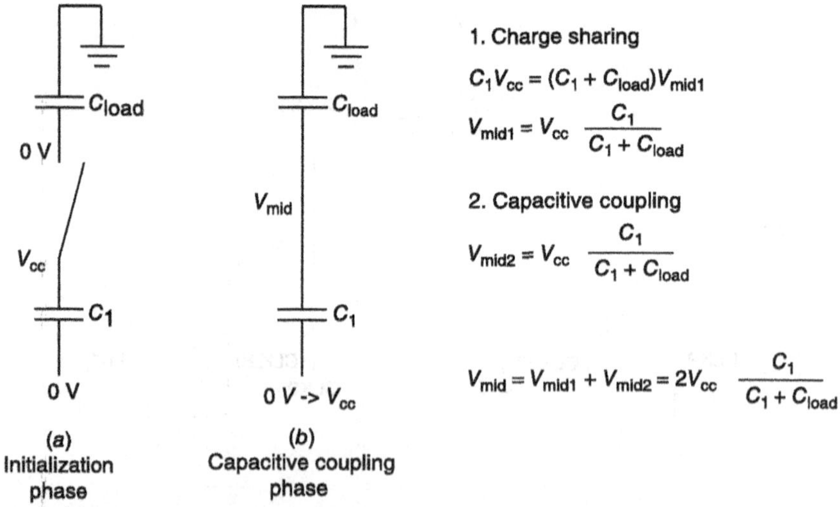

Figure 5.13 Effective coupling voltage of clock doubler.

superposition theorem. The final potential across plates of C_{load} could be calculated by Eq. (5.29).

$$V_{mid} = 2V_{cc}\frac{C_1}{C_1 + C_{load}} \tag{5.29}$$

There are three possible scenarios based on the relationship between C_1 and C_{load}. First, if $C_1 >> C_{load}$, then V_{mid} could be approximated to $2V_{cc}$. This is the best case for coupling. Second, if $C_1 = C_{load}$, then V_{mid} will be equal to V_{cc}. Third, if $C_1 << C_{load}$, then V_{mid} will be near 0. Only the first scenario makes sense from an efficiency point of view.

In real design, the effective capacitance C_1 should be much greater than C_{load} to maximize the benefit of using the pump clock amplitude doubler.

Figure 5.14 illustrates the normalized effective voltage versus the capacitance coupling ratio. If the capacitive ratio between C_1 and C_{load} could reach 5, effectively the coupling voltage could reach nearly 1.7*x* of V_{cc}. If C_1 and C_{load} were of the same device, the area efficiency would be very low in order to achieve such high coupling ratio. In real design, normally C_1 and C_{load} are made of different types of capacitors. Capacitor C_1 is normally made of a thinner oxide device with large unit capacitance per unit area, while C_{load} has to withstand higher voltage and is made of thicker oxide device. Using this approach enables maximization of both power efficiency and area efficiency at the same time in the design at charge pump system level.

5.3.1.2 Extra Clock Phase

When two capacitors are cascaded in series in the clock doubler, the effective capacitance is reduced in design. This approach will

C_1/C_{load}	V_{mid}/V_{cc}
1	1.00
2	1.33
3	1.50
4	1.60
5	1.67
6	1.71
7	1.75
8	1.78
9	1.80
10	1.82

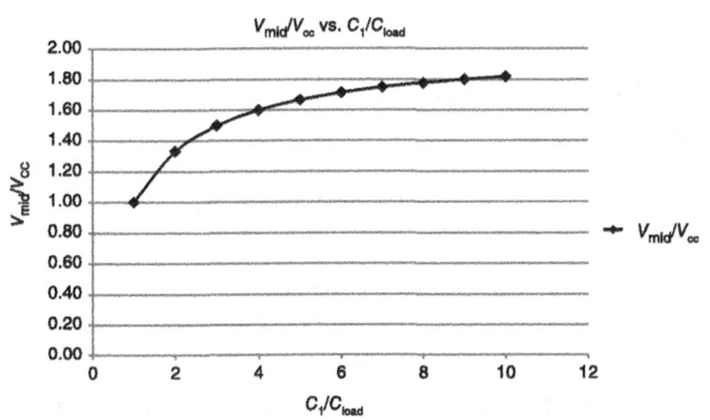

FIGURE 5.14 Normalized effective voltage versus capacitive coupling ratio.

hurt the area efficiency of the capacitors. How can one improve the effective capacitance in this type of design?

Figure 5.15 illustrates a design approach[2] to use an extra clock phase to improve the efficiency of capacitance. There are a total of three steps for capacitive coupling: the initialization phase, precharging phase, and capacitive coupling phase.

First, the initialization phase is the same as the normal approach illustrated in Fig. 5.13*a*. The charge stored in C_1 is equal to C_1V_{cc}. The charge stored on C_{load} is 0.

Second, the added precharging phase charges up the bottom plate of C_{load}. At the end of this phase, the charge stored in C_{load} is $C_{load}V_{cc}$.

After the first two phases, Eq. (5.30) gives the effective voltage V_{mid} after the switch connects the bottom plate of C_{load} and top plate of C_1.

$$V_{mid} = V_{cc}\frac{C_1 + C_{load}}{C_1 + C_{load}} = V_{cc} \tag{5.30}$$

Third, after the final clock boost from 0 to V_{cc}, the effective driving voltage V_{mid} on C_{load} is given by Eq. (5.31).

$$V_{mid} = V_{cc}\left(1 + \frac{C_1}{C_1 + C_{load}}\right) \tag{5.31}$$

With the added clocking phase, the effective gain on the driving voltage on C_{load} has been increased compared to Eq. (5.29).

Figure 5.16 plots the normalized results of the effective driving strength as a function of the ratio of C_1 to C_{load}. Both old and new designs are calculated and plotted for comparison. For a high coupling ratio, the improvement is about 10%. The improvement is maximized at a lower capacitive ratio. There is nearly a 20% improvement

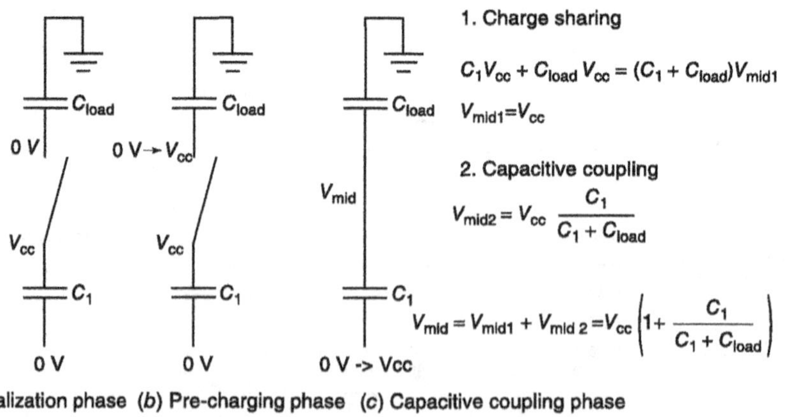

FIGURE 5.15 Extra clock phase to improve capacitance efficiency.

C_1/C_{load}	V_{mid}/V_{cc} (New)	V_{mid}/V_{cc} (Old)
1	1.50	1.00
2	1.67	1.33
3	1.75	1.50
4	1.80	1.60
5	1.83	1.67
6	1.86	1.71
7	1.88	1.75
8	1.89	1.78
9	1.90	1.80
10	1.91	1.82

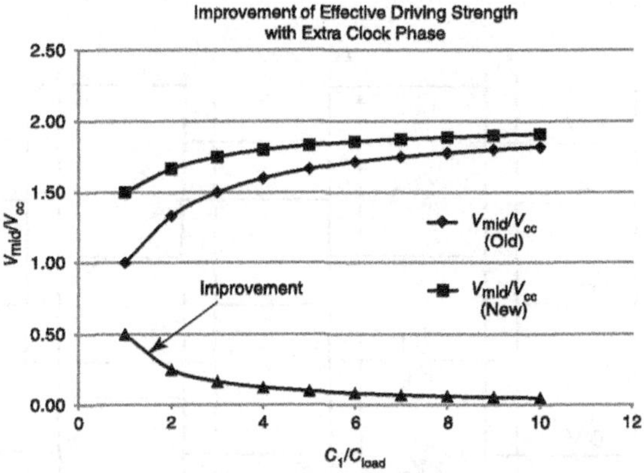

FIGURE 5.16 Improvement of effective driving strength with extra clock phase.

in the effective driving strength with a coupling ratio between 4 and 5. With coupling ratio of 1, the improvement is 50%.

5.3.1.3 Four-Phase Pump Clock Doubler

Figure 5.17 illustrates the four-phase pump clock doubler.[2] It is modified from the original design in Fig. 5.10 to accommodate additional clock phases from CLK1G and CLK2G. The four-phase pump clock doubler modifies the clock gate signal to allow the precharge of load with V_{cc} before final clocking from the main clocks.

Figure 5.18 illustrates the clock waveforms from the four-phase pump clock doubler. CLK1 and CLK2 are input clocks that drive the

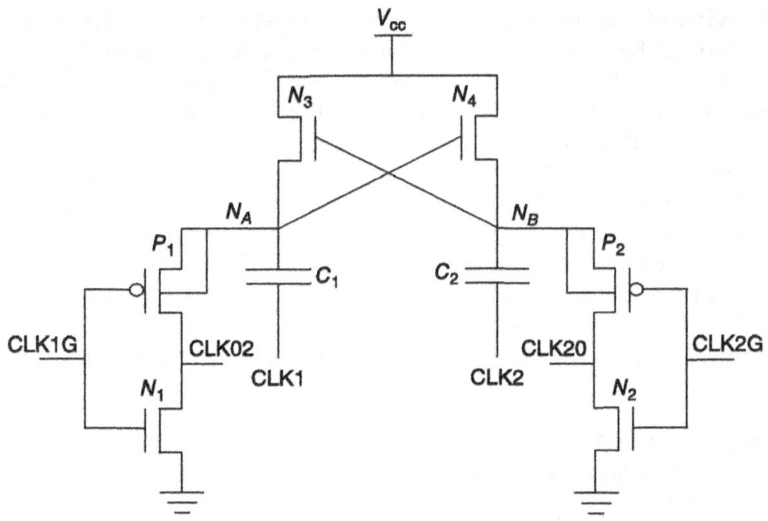

FIGURE 5.17 Four-phases pump clock doubler.

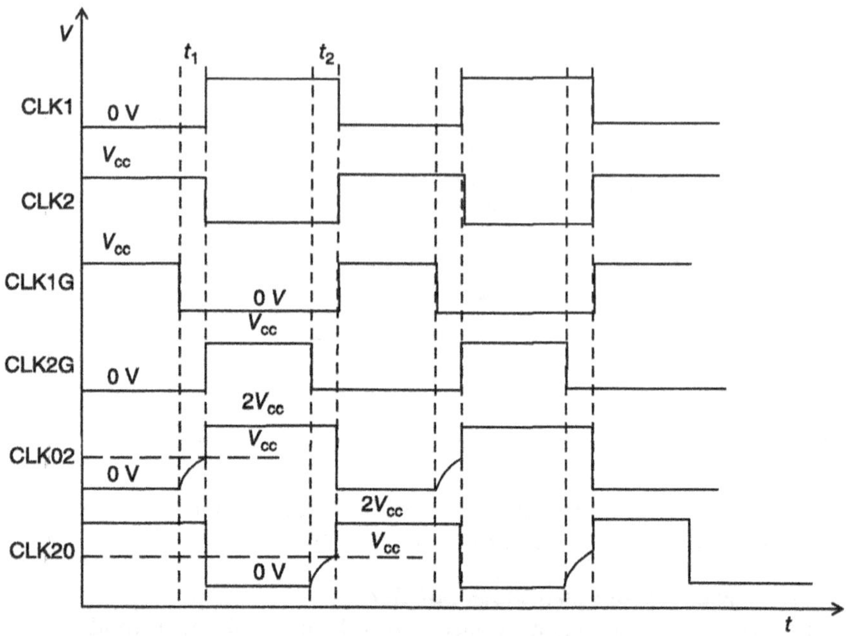

FIGURE 5.18 Clock waveforms of four-phases clock doubler.

coupling capacitance. They should not overlap in high clock phases. CLK1G is opposite in phase from CLK1 to pass the internal charge from *NA*. To precharge the load, the falling edge of CLK1G is pulled in from the rising edge of CLK1 by the duration of t_1 to allow pre-charging of the output from V_{cc} with the existing circuits. The rising edge of CLK1G is opposite in phase to the falling edge of CLK1 to discharge the CLK02, and to cut off the path between N_A and CLK02. Similar reasoning could be applied to the clock phase relationship between CLK2G and CLK2.

In Fig. 5.18, CLK02 has a two-step ramp up in the rising edge transition. Initially CLK02 will rise to V_{cc} through the same reset path that reset node N_A back to V_{cc}. After a sufficient duration t_1, CLK02 is AC coupled by the rising edge of CLK1. Then CLK02 will fall following the falling edge of CLK1. The waveform of CLK20 is similar to CLK02 with a shifted phase. At the end, CLK02 and CLK20 with doubled clock amplitude are generated with improved coupling efficiency.

5.4 References

1. Pan, F. "Analog Assisted DSR." (Engineering Notes and patent application is in process)
2. Pan, F., J. H. Huynh. "Charge Pump with Reduction in Inefficiencies Due to Charge Sharing between Capacitances. U.S. patent no. 8,421,524 April 2013."

CHAPTER 6

Charge Pump Stability Analysis

The closed-loop regulation of the charge pump relies on negative feedback to achieve control over the gain and linearity. Similar to many other analog systems with negative feedback, such as operational amplifiers (op-amps), the stability of charge pump system with negative feedback must be analyzed. Proper compensation should be applied to ensure the stability of design.

6.1 Negative Feedback

With a known reference voltage and large open-loop gain, negative feedback determines the closed-loop gain of the charge pump. Characteristics of active components could vary in large degree with respect to the supply voltage, temperature, process, etc. Characteristics of passive elements have much smaller variation. With better matching and less variations in design, using only passive elements such as resistors or capacitors in the feedback network, the closed-loop gain, linearity, and bandwidth could be well controlled.[1–3]

With negative feedback, the returned signal would be forced by large open-loop gain to be near the reference target. With the minimized error term between the returned signal and reference, the linearity and distortion of the system are improved.[2] The closed-loop gain is set by the feedback ratio rather than the device parameters of the active components. The higher the open-loop gain that the charge pump system could achieve, the lesser the output error would be with respect to PVT corners.

Figure 6.1 illustrates a generic representation of a negative feedback system, where $V_{in}(s)$ is the input signal, $V_{out}(s)$ is the output signal of the system, $A(s)$ is defined as the forward gain, and $\beta(s)$ is the feedback ratio. All the terms are in the Laplace transform for the ease of frequency domain analysis.

$$V_{out}(s) = [V_{in}(s) - \beta(s)V_{out}(s)]A(s) \tag{6.1}$$

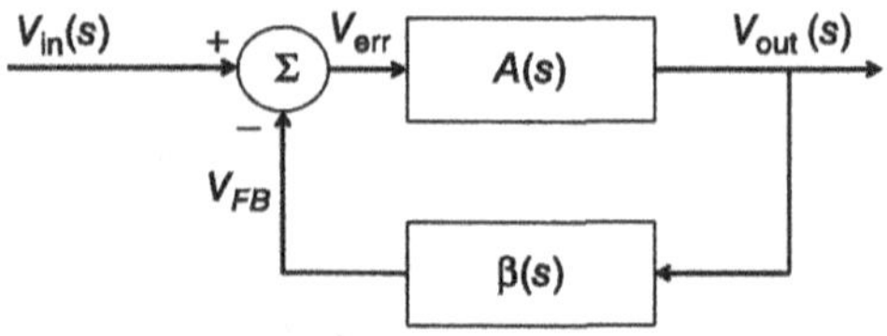

FIGURE 6.1 Negative feedback system.

$$H(s) = \frac{V_{out}(s)}{V_{in}(s)} = \frac{A(s)}{1 + A(s)\beta(s)} \tag{6.2}$$

$V_{out}(s)$ the closed-loop response of system with input stimulus $V_{in}(s)$ could be derived from Eq. (6.1). Equation (6.2) determines the overall system transfer function with negative feedback.

$$A_{OL}(s) = A(s)\beta(s) \tag{6.3}$$

$$L(s) = A(s)\beta(s) = A_{OL}(s) \tag{6.4}$$

Equation (6.3) defines the open-loop gain $A_{OL}(s)$ of the system over the entire frequency range. Equation (6.4) defines the loop gain $L(s)$ of given system which is same as $A_{OL}(s)$.

$$H(s) = A_{CL}(s) = \frac{A(s)}{1 + A_{OL}(s)} \tag{6.5}$$

$$H(s) \approx \frac{1}{\beta(s)} \quad \text{if} \quad L(s) >> 1 \tag{6.6}$$

Equation (6.5) defines the transfer function $H(s)$ or closed-loop gain $A_{CL}(s)$ of the given system. As long as $L(s)$ is much greater than 1, $A_{CL}(s)$ could be approximated by Eq. (6.6) with dependence on $\beta(s)$ only.

$$\frac{\partial H(s)}{\partial A(s)} = \frac{1}{1 + [A(s)\beta(s)]^2} \approx \frac{1}{L(s)^2} \tag{6.7}$$

If the sensitivity of $H(s)$ is taken with respect to $A(s)$, as shown in Eq. (6.7), with negative feedback, the sensitivity to $A(s)$ is suppressed by loop gain $L(s)$ if $L(s) >> 1$.

6.2 Stability

In order for the negative feedback system to be stable the bounded input, it must generate bounded output. If $A_{CL}(s)$ approaches infinity at a certain frequency, the output of the system will not be bounded with a smaller change of input or small perturbation of noise.

This situation occurs if the denominator of Eq. (6.5) approaches or equals zero at a certain frequency. Any small disturbance to the negative feedback loop will generate amplified return, which in turn will be further amplified through the closed-loop and oscillation will start. In this situation, negative feedback turns into positive feedback at the given frequency, and an intended stable system with negative feedback becomes unstable at given high frequency.

6.2.1 DC or Low-Frequency Response

With negative feedback control, the closed-loop gain of the system at DC or low frequency could be simplified into Eq. (6.8).

$$A_{CL}(0) = \frac{A(0)}{1 + A_{OL}(0)} \tag{6.8}$$

Then $A(s)$ and $A_{OL}(s)$ are evaluated with $s = 0$.

$$A_{CL}(0) \approx \frac{1}{\beta(0)} \quad \text{if} \quad A_{OL}(0) >> 1 \tag{6.9}$$

The amplitude of the DC open-loop gain $|A_{OL}(0)|$ is much greater than 1. The total phase shift at DC or low frequency across the loop is near zero. Equation (6.6) could be approximated by Eq. (6.9). At DC or low frequency, the negative feedback system is stable with the closed-loop gain defined by the feedback ratio $\beta(0)$. Forward gain $A(s)$ depends on the characteristics of many active components, such as BJTs or MOSFETs. It is subject to large variations over PVT corners. With closed-loop gain defined by $\beta(0)$, a higher degree of accuracy is achieved and sensitivity to variation of $A(s)$ is suppressed.

6.2.2 High-Frequency Response

As the frequency of the signal increases, negative feedback could be turned into positive feedback with the sign change of $A_{OL}(s)$ in denominator of Eq. (6.5).

$$1 + A_{OL}(\omega_0) = 0 \tag{6.10}$$

$$H(s) = \frac{A(\omega_0)}{0} = \infty \tag{6.11}$$

This situation would occur if Eq. (6.10) held true at frequency ω_0. The closed-loop gain of the system at a given frequency ω_0 is given by Eq. (6.11). It reaches infinity, and any disturbance to the loop would cause positive feedback and eventually rail-to-rail oscillation.

To investigate the stability of the negative feedback loop, open-loop Bode plots could be analyzed. Both magnitude and phase plots of loop gain $L(s)$ are plotted over the entire frequency range. If the phase of loop gain $\angle L(\omega)$ does not change 180° before magnitude $|L(\omega)|$ reaches 0 dB, then the system will be stable.

$$PM = \angle L(\omega_t) + 180° \tag{6.12}$$

The phase margin (PM) is defined as the additional phase shift that would be required at unity-gain frequency to cause instability. The definition is given by Eq. (6.12). Generally large phase margins between 45° and 60° are required. A smaller phase margin leads to ringing in dynamic settling behavior.

6.2.3 First-Order Negative Feedback System

The first-order negative feedback system with a dominant pole could be described by Eq. (6.13):

$$L(s) = \frac{L_0}{1 + s/\omega_{p1}} \tag{6.13}$$

At DC or very low frequency with $\omega << \omega_{p1}$, the open-loop gain and phase shift could be approximated by Eqs. (6.14) and (6.15).

$$|L(\omega)| \approx L_0 \quad \text{for} \quad \omega << \omega_{p1} \tag{6.14}$$

$$\angle L(\omega) \approx 0° \quad \text{for} \quad \omega << \omega_{p1} \tag{6.15}$$

At very high frequency with $\omega >> \omega_{p1}$, the open-loop gain and phase shift could be approximated by Eqs. (6.16) and (6.17).

$$|L(\omega)| \approx \frac{L_o}{\omega/\omega_{p1}} \quad \text{for} \quad \omega >> \omega_{p1} \tag{6.16}$$

$$\angle L(\omega) \approx -90° \quad \text{for} \quad \omega >> \omega_{p1} \tag{6.17}$$

At frequency $\omega = \omega_{p1}$, the open-loop gain and phase shift could be approximated by Eqs. (6.18) and (6.19).

$$|L(\omega)| = \frac{L_0}{\sqrt{2}} \quad \text{at} \quad \omega = \omega_{p1} \tag{6.18}$$

$$\angle L(\omega) = -45° \quad \text{at} \quad \omega = \omega_{p1} \tag{6.19}$$

With $\omega << \omega_{p1}$, the magnitude of the loop gain is near constant dc gain L_0, and the loop phase shift is near 0°. At $\omega = \omega_{p1}$, the magnitude of the loop gain drops 3 dB from dc gain L_0, open-loop phase shifts −45°. At higher frequency, with $\omega >> \omega_{p1}$, the magnitude of the loop gain decays at a rate of 20 dB/decade, and the loop phase shifts near −90°. With a single pole system, the maximum phase shift is −90° and it is unconditionally stable.

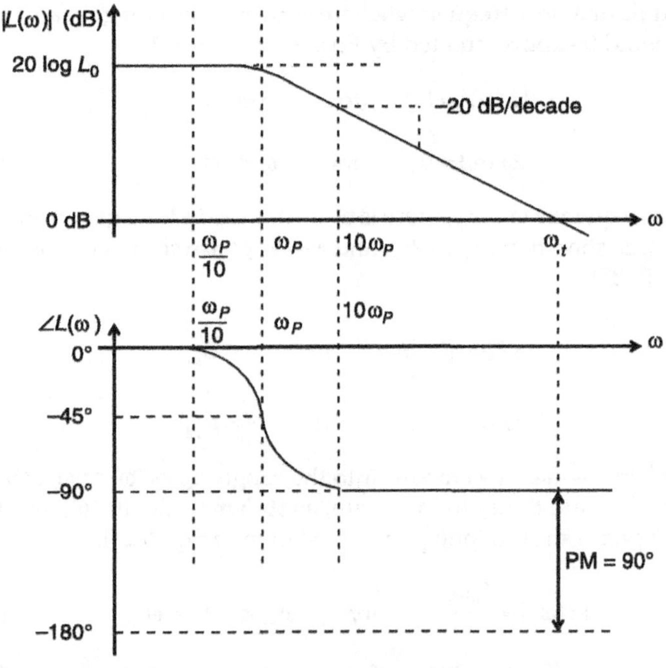

FIGURE 6.2 Bode plot of first-order negative feedback system.

Figure 6.2 illustrates the Bode plot of the first-order negative feedback system. The dominant pole is located at ω_p, and $|L(\omega_p)|$ is –3 dB from its dc gain. As the Bode plot shows, the pole has phase impact one decade above the pole frequency and one decade below the pole frequency.

$$\omega_t = L_0\omega_p \tag{6.20}$$

The *unity-gain frequency* ω_t is defined as the frequency at which $|L(\omega_t)| = 1$. The unity-gain frequency ω_t could be derived from Eq. (6.16) and it is defined in Eq. (6.20). The phase margin of the first-order negative feedback system is 90°.

6.2.4 Second-Order Negative Feedback System

Although the first-order feedback system has many great benefits, in practice, the negative feedback system usually has multiple poles within or near the bandwidth of interest. In many situations, a complicated system could be approximated by a second-order system with or without compensation. The transfer function for a general second-order feedback system, with the assumption $\omega_{p1} \ll \omega_{p2}$, is

$$L(s) = \frac{L_0}{(1+s/\omega_{p1})(1+s/\omega_{p2})} \tag{6.21}$$

At dc or at low frequency where $\omega << \omega_{p1}$, the loop gain and phase shift could be approximated by Eqs. (6.22) and (6.23).

$$|L(\omega)| \approx L_0 \quad \text{for} \quad \omega << \omega_{p1} \tag{6.22}$$

$$\angle L(\omega) \approx 0° \quad \text{for} \quad \omega << \omega_{p1} \tag{6.23}$$

At frequency $\omega = \omega_{p1}$, there is a –3-dB drop in loop gain from its dc level L_0 as shown in Eq. (6.24), and a –45° phase shift at ω_{p1}, as shown in Eq. (6.25).

$$|L(\omega)| \approx \frac{L_o}{\sqrt{2}} \quad \text{at} \quad \omega = \omega_{p1} \tag{6.24}$$

$$\angle L(\omega) \approx -45° \quad \text{at} \quad \omega = \omega_{p1} \tag{6.25}$$

When frequency ω moves into the range $\omega_{p1} << \omega << \omega_{p2}$, there is –20 dB/decade decay in loop gain, as shown in Eq. (6.26), and there is –90° phase shift in loop phase, as shown in Eq. (6.27).

$$|L(\omega)| \approx \frac{L_0 \omega_{p1}}{\omega} \quad \text{for} \quad \omega_{p1} << \omega << \omega_{p2} \tag{6.26}$$

$$\angle L(\omega) \approx -90° \quad \text{for} \quad \omega_{p1} << \omega << \omega_{p2} \tag{6.27}$$

At frequency $\omega = \omega_{p2}$, Eq. (6.28) defines the loop gain at the given frequency ω_{p2}, and Eq. (6.29) shows there is a –135° phase shift at the second pole frequency.

$$|L(\omega_{p2})| \approx \frac{\omega_{p1}}{\omega_{p2}} \frac{L_o}{\sqrt{2}} \quad \text{at} \quad \omega = \omega_{p2} \tag{6.28}$$

$$\angle L(\omega_{p2}) \approx -135° \quad \text{at} \quad \omega = \omega_{p2} \tag{6.29}$$

At a frequency much greater than the second pole frequency, $\omega >> \omega_{p2}$, the loop gain decays at rate of –40 dB/decade, as shown in Eq. (6.30), and there is a –180° phase shift in loop phase, as shown in Eq. (6.31).

$$|L(\omega)| = L_0 \frac{\omega_{p1}\omega_{p2}}{\omega^2} \quad \text{for} \quad \omega >> \omega_{p2} \tag{6.30}$$

$$\angle L(\omega) \approx -180° \quad \text{for} \quad \omega >> \omega_{p2} \tag{6.31}$$

The second-order system has the potential to shift –180° in loop phase. Proper compensation is needed to ensure a large phase margin for fast settling and good transient response.

Figure 6.3 illustrates the Bode plot of a second-order negative feedback system. The two poles ω_{p1} and ω_{p2} are separated relatively far away in the frequency domain. As the frequency moves higher,

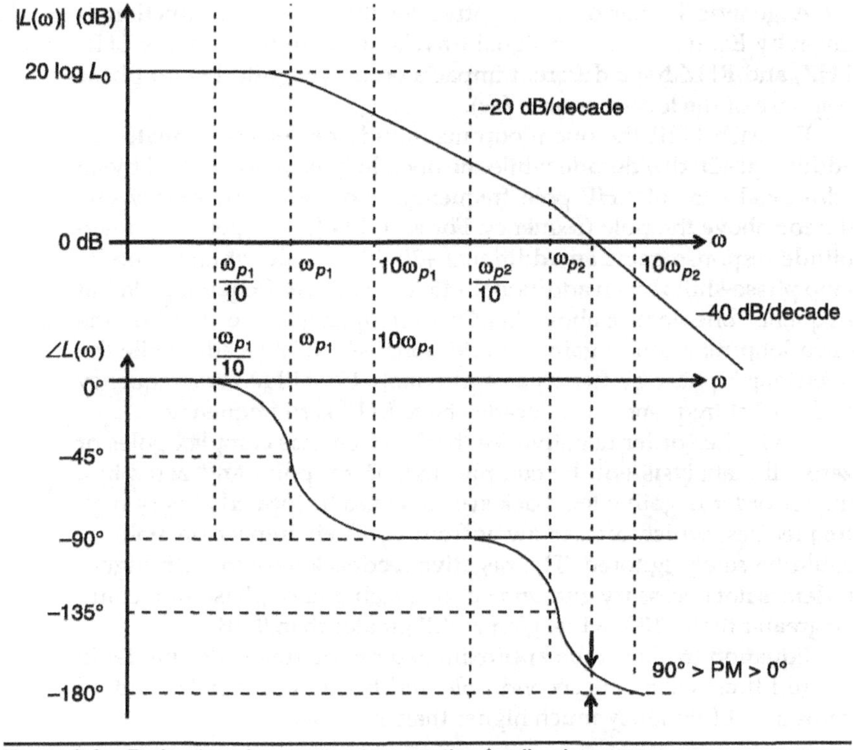

FIGURE 6.3 Bode plot of second-order negative feedback system.

each pole contributes to –20 dB/decade of decay in the loop gain. There is maximum –40 dB/decade of decay after the frequency passes two poles. Each pole can contribute maximum –90° phase shift, and there is maximum –180° phase shift in loop phase. At each pole, there is –45° phase shift.

In the illustration, ω_{p1} and ω_{p2} have been separated far enough. At the frequency where the loop gain reaches 0 dB, the open-loop phase shift has not reached –180° yet in the system.

6.2.5 Higher-Order Negative Feedback System

In a practical negative feedback system, the signal could travel through many nodes in the loop. There could be many paths between two nodes in the loop where the signal could travel. The transfer function of a practical negative feedback system can contain additional left half-plane (LHP) poles, left half-plane zeros (LHZ), right half-plane zeros (RHZ), etc.

$$A(s) = A_0 \frac{(1+s/\omega_{z1})(1+s/\omega_{z2})\cdots(1+s/\omega_{zm})}{(1+s/\omega_{p1})(1+s/\omega_{p2})\cdots(1+s/\omega_{pn})} \tag{6.32}$$

A generic higher-order negative feedback transfer function is given by Eq. (6.32). As the signal travels through the loop, the LHP, LHZ, and RHZ have different impacts on the magnitude and phase response of the loop response $L(s)$.

For each LHP, the open-loop magnitude response attenuates an additional –20 dB/decade while the open loop is phase-shifted by an additional –45° at LHP pole frequency, and –90° at frequency one decade above the pole frequency. For each LHZ, the open-loop magnitude response gains an additional +20 dB/decade while the open-loop phase-shifts by an additional +45° at LHZ zero frequency, +90° at frequency one decade above LHZ zero frequency. For each RHZ, the open-loop magnitude gains an additional +20 dB/decade while the open loop is phase-shifted by an additional –45° at RHZ zero frequency and –90° at frequency one decade above RHZ zero frequency.

If a higher-order negative feedback system has complex poles or zeros, the analysis will be complicated. Many poles and zeros in a higher-order negative feedback system could be located at very high frequencies, which are far away from ω_t. Their impact on stability could be safely ignored. The negative feedback system with higher orders is not necessary guaranteed to be stable since phase shift could be greater than –180° while gain is still greater than 0 dB.

Equation (6.32) can be approximated by the first-order model in Eq. (6.13) on when the second pole and higher-order poles and all zeros are at frequency much higher than ω_t.

$$\omega_t << |\omega_{p2}|, \ldots, |\omega_{pn}|, |\omega_{z1}|, \ldots, |\omega_{zm}| \tag{6.33}$$

Higher-order systems can be approximated by the second-order model in Eq. (6.21) with approximation. The zero-value time-constant technique[1] can be used to estimate the bandwidth of system with Eq. (6.34).

$$\frac{1}{\omega_{pd}} \cong \sum_{i=1}^{n} \frac{1}{\omega_{pi}} \tag{6.34}$$

A second pole can approximate all higher-frequency poles and zeros with Eq. (6.35).[2]

$$\frac{1}{\omega_{eq}} \cong \sum_{i=2}^{n} \frac{1}{\omega_{pi}} - \sum_{i=1}^{m} \frac{1}{\omega_{zi}} \tag{6.35}$$

6.2.6 Compensation

The goal of compensation is to create one effective dominant pole directly or indirectly below ω_t in the open-loop transfer function $L(s)$ and to ensure a reasonably large positive phase margin for stability and dynamic settling.

If the system can approximate a first-order negative feedback system after compensation as in Eq. (6.13), it will be unconditionally stable with near 90° phase margin. If the compensated system can approximate a second-order model as in Eq. (6.21), a nondominant pole is allowed at frequency of ω_t or at high frequency above ω_t. LHZ can be positioned slightly above ω_t to improve the phase margin of the loop response.

Compensation could be applied in design to the forward gain amplifier $A(s)$, loading network, or feedback network $\beta(s)$. The compensation manipulates the relationship between loop gain $|L(s)|$ and loop phase $\angle L(s)$ with the addition of poles or zeros.

There are a few common techniques for compensation, such as dominant pole compensation, lead compensation, and pole splitting technique.

Dominant pole compensation is one of most commonly used techniques. The dominant pole frequency is positioned low enough to force the open-loop response $L(s)$ to act as a single-pole system for frequency below and near ω_t. This is a form of lag compensation. High-frequency poles and zeros should be positioned above ω_t. Decreased unity-gain bandwidth of $L(s)$ is expected.

Figure 6.4 illustrates the Bode plots of dominant pole compensation. There are two poles of interest in open-loop response $L(s)$

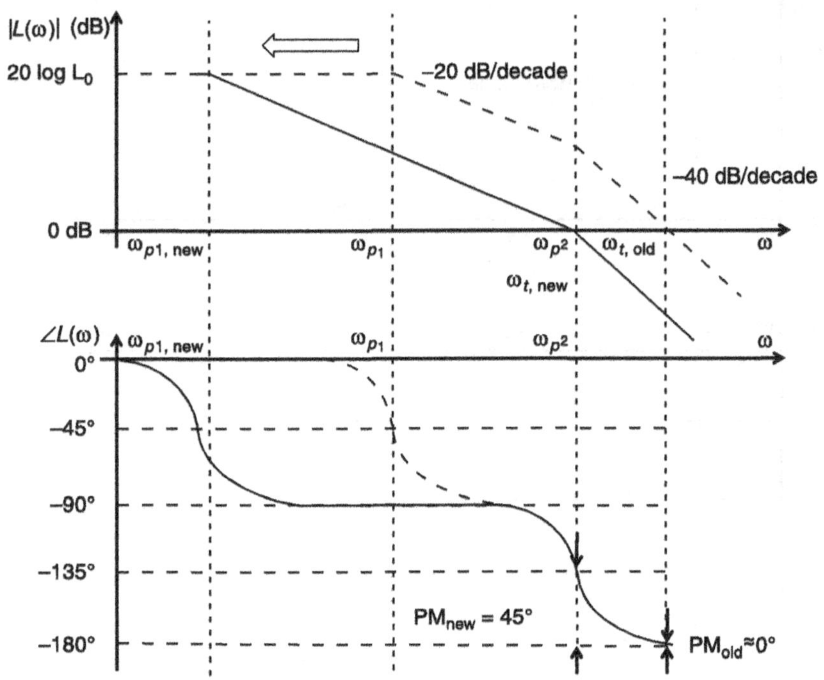

Figure 6.4 Bode plots of dominant pole compensation.

before compensation. Poles frequencies are not separated far apart, and phase margin is near 0°. With dominant pole compensation, dominant pole frequency is set to be at $\omega_{p1,\,new}$. New unity-gain frequency $\omega_{t,\,new}$ coincides with the nondominant pole ω_{p2} and creates 45° phase margin after compensation.

Lead compensation technique[2-3] can further increase the phase margin with the addition of left half-plane ω_z at frequency slightly above ω_{p2}.

Figure 6.5 illustrates Bode plots with additional lead compensation. There is little change in the unity-gain frequency, but there is an additional +20° phase margin improvement as highlighted by the solid line with lead compensation.

Miller-cap compensation[1-3] is another popular technique for compensation with moderate load period.[4] Pole splitting allows the creation of a dominant pole at lower frequency and pushes the second nondominant pole to higher frequency. Compared with the dominant pole compensation technique, the unity-gain bandwidth with Miller-cap compensation would be higher. A nulling resistor could be added in series with Miller capacitance to cancel RHZ, or convert RHZ to LHZ, to further improve the phase margin.

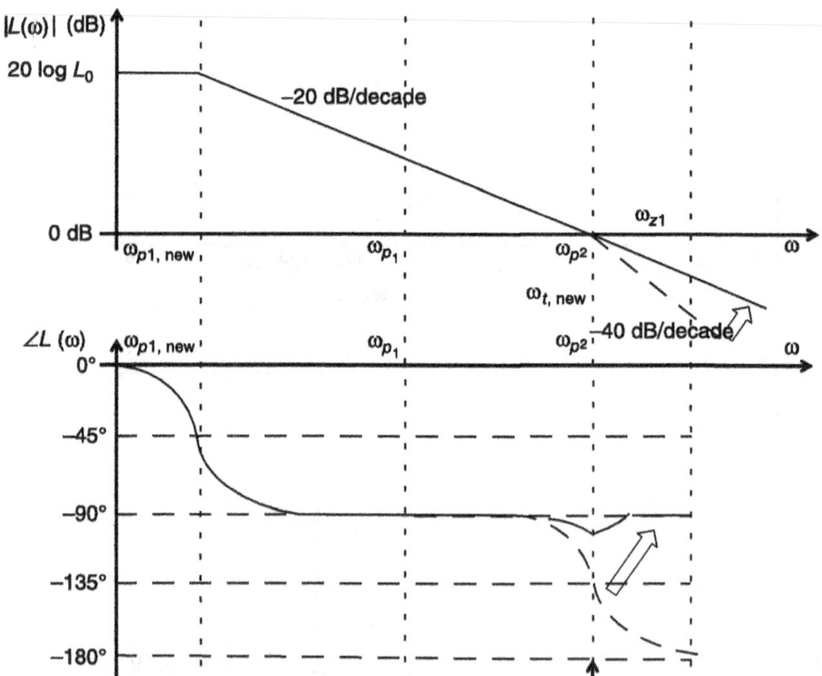

Figure 6.5 Bode plots with additional lead compensation.

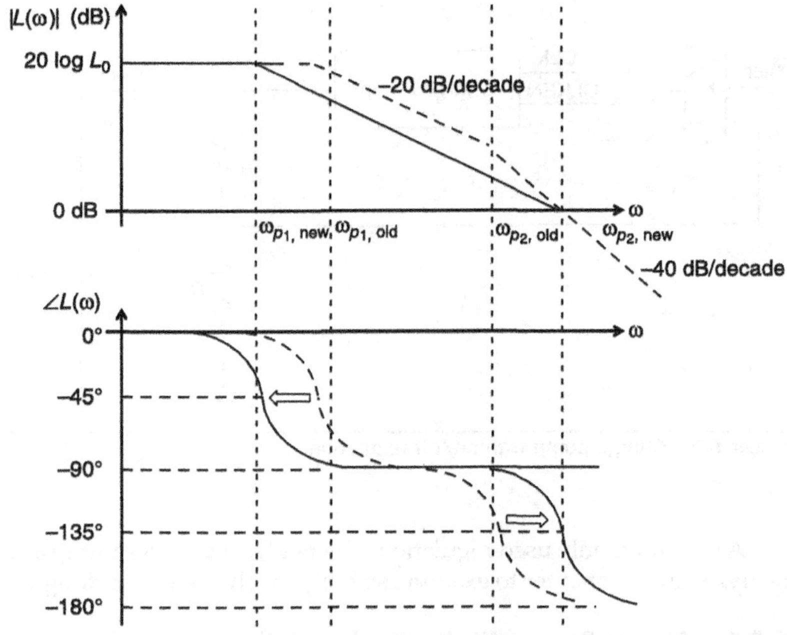

FIGURE 6.6 Bode plots for Miller-cap compensation.

Figure 6.6 illustrates Bode plots for Miller-cap compensation. The dotted line represents open-loop response $L(s)$ before compensation. The solid line represents the system after Miller-cap compensation. Low-frequency pole $\omega_{p1,\,old}$ moves to even lower frequency at $\omega_{p1,\,new}$, and nondominant pole $\omega_{p2,\,old}$ moves to higher frequency at $\omega_{p2,\,new}$. The unity-gain frequency of the compensated system coincides with $\omega_{p2,\,new}$. Phase margin improves to 45° with Miller-cap compensation.

Besides these well-known techniques, many other compensation methods could be used to achieve good phase margin and stability with reasonable efforts.

6.3 Stability and Compensation for Charge Pump

There are challenges associated with stability analysis of the charge pump. Unlike op-amp or other analog designs, small-signal AC analysis cannot be performed with Spice on the charge pump system owing to the switching nature of design. The small-signal model for components in the charge pump requires switching to establish operating points, which could not be achieved with DC biases. With charge pump design, only transient simulations could be used to validate and analyze the stability of design. In addition, the difficulty of differentiating the stability issue from switching noise is another challenge.

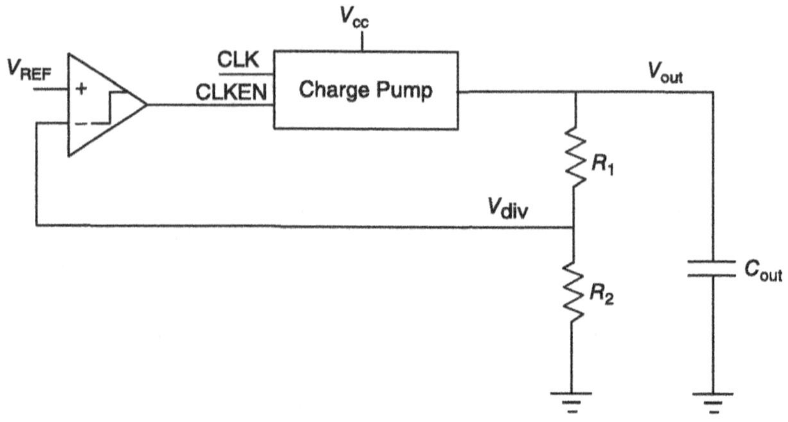

FIGURE 6.7 Charge pump with on/off regulation.

A few commonly used regulation schemes for the charge pump are analyzed in this chapter to examine stability for charge pump designs.

6.3.1 Charge Pump with On/Off Regulation

Figure 6.7 illustrates the charge pump with on/off regulation. This is the most commonly used regulation scheme for charge pump design. Enabling or disabling of the charge pump clocks is controlled by comparator output CLKEN and V_{REF} is a known reference voltage.

Is the architecture illustrated in Fig. 6.7 a stable system or an unstable system? The charge pump with on/off regulation has similar topology to the op-amp with negative feedback, except the comparator is used in the on/off regulation scheme. With the comparator in the feedback loop, the output is quantized based upon the regulation level. If V_{out} exceeded the target level, the charge pump would be disabled. If V_{out} fell below the target level, the charge pump would be enabled. Using the comparator to enable or disable the negative feedback loop, the charge pump with an on/off regulation scheme is unconditionally stable. A stable system does not mean the output noise would be small if proper care is not taken.

Figure 6.8 illustrates the output noise of the charge pump with on/off regulation. There are two different designs but with identical target regulation levels V_{target}. From the transient waveforms, both (*a*) and (*b*) are stable since outputs are bounded.

In Fig. 6.8*a*, the transient waveform is bounded between V_{nt1} and V_{nb1}. The maximum output noise is confined within ΔV_{noise1}. For the rising edge, there are about three pump clock cycles or clock edges to bring the output potential above V_{target} before the pump clock is disabled. The load current and feedback current determine the slew rate of the falling edge. Similarly in Fig. 6.8*b*, the transient waveform is

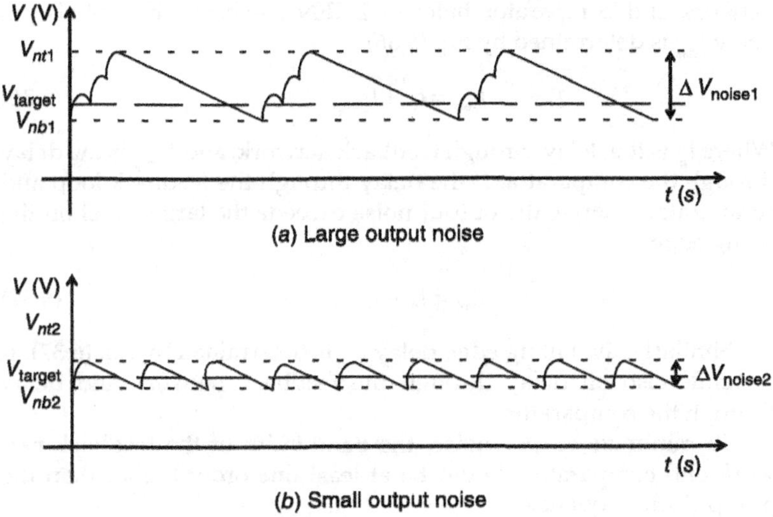

Figure 6.8 Output noises of charge pump with on/off regulation.

bounded between V_{nt2} and V_{nb2}. The maximum noise of the waveform is confined within ΔV_{noise2}. What could cause the difference between case (*a*) and case (*b*) in Fig. 6.8?

Although the on/off regulation scheme is unconditionally stable, the bandwidth of the open-loop response determines the output noise and power consumption of design.

Figure 6.9 illustrates noises versus delays in charge pump design. Figure 6.9*a* shows that it takes the delays of t_{rise} for the comparator to switch and turn off charge pump and t_{fall} for the comparator to switch and turn off the charge pump. For the rising edge delay, as V_{out} rises above V_{target}, the output will traverse the resistive divider feedback

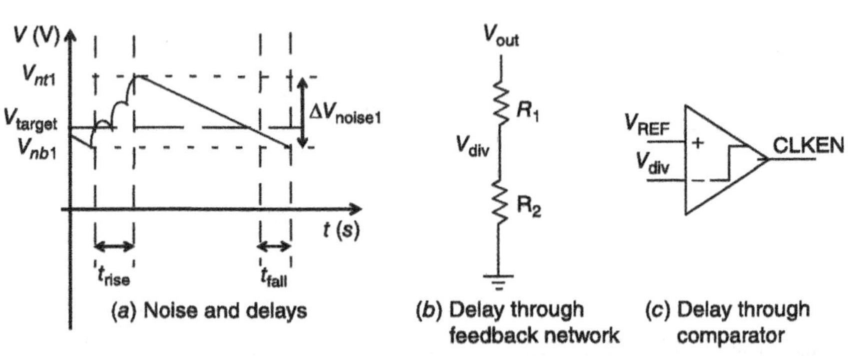

Figure 6.9 Noises versus delays.

network and comparator before CLKEN switches off. The rise in delay t_{rise} is determined by Eq. (6.36).

$$t_{rise} = t_{fb} + t_{comp} \tag{6.36}$$

Where t_{fb} is the delay through feedback network, and t_{comp} is the delay through the comparator. If the delay through the feedback loop and comparator is large, the output noise exceeds the target level on the rising edge.

$$t_{fall} = t_{fb} + t_{comp} \tag{6.37}$$

Similarly, the falling edge delay t_{fall} is determined by Eq. (6.37). It depends also on delay through the feedback network and delay through the comparator.

To minimize output noise, the bandwidth of the feedback network and comparator should be at least one order higher than the pump clock frequency.

6.3.2 Charge Pump with Dynamic Voltage Scaling

Figure 6.10 illustrates charge pump regulation with dynamic voltage scaling on the pump clock amplitude. The charge pump clock driver is enabled from the beginning of operation without interruption. Unlike the on/off regulation scheme, the pump clock runs at fixed frequency. If V_{out} deviates from target, the o-amp will move $V_{CLK,SUP}$ up or down to oppose the change of V_{out}.

Figure 6.11 illustrates the regulation of power supply for pump clock drivers. The difference between V_{div} and V_{REF} is amplified, and the output of the error amplifier is *NP*, which drives the gate of the

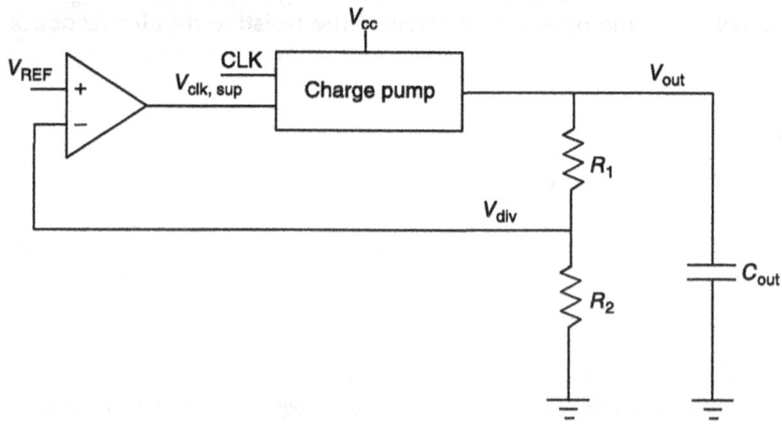

FIGURE 6.10 Regulation with dynamic voltage scaling on pump clock amplitude.

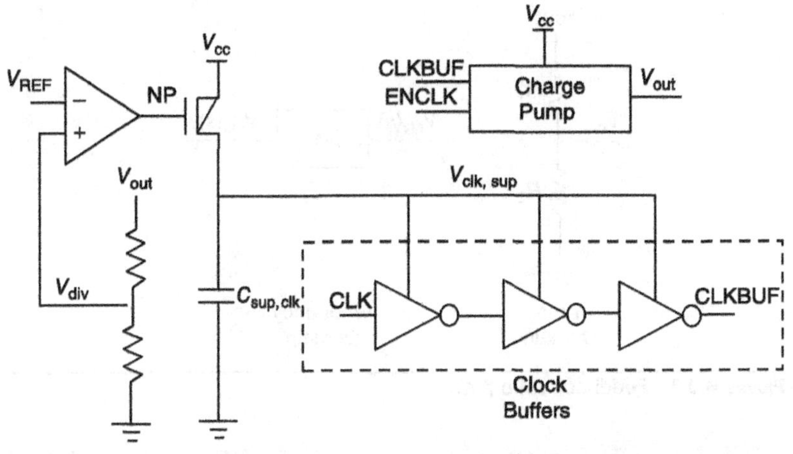

FIGURE 6.11 Regulation of power supply for pump clock drivers.

PMOS power switch. Voltage $V_{clk,sup}$ is the power supply for pump clock drivers. The buffered clock signal CLKBUF swings between 0 V and $V_{clk,sup}$. Dynamic voltage scaling intends to minimize the output noise against PVT variation, or line or load variation with moderate decoupling capacitance $C_{sup,clk}$.

With negative feedback regulation of the pump clock amplitude, stability analysis should be performed. It is natural to understand the design from small-signal modeling and then be able to analyze stability and compensation from transient waveforms.

6.3.2.1 Charge Pump Small-Signal Model

Figure 6.12 illustrates the small-signal model for pump clock amplitude regulation. The transfer function $V_{out}(s)$ could be derived from Eq. (6.38), where $A_{amp}(s)$ is the transfer function of the error amplifier, $A_{pump}(s)$ is the transfer function associated with charge pump, and $\beta(s)$ is the feedback network for sampling.

$$V_{out}(s) = \frac{A_{amp}(s)A_{pump}(s)}{1 + A_{amp}(s)A_{pump}(s)\beta(s)} V_{REF} \tag{6.38}$$

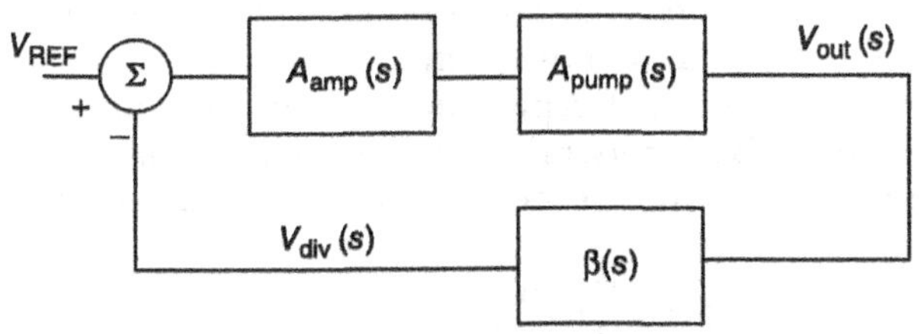

FIGURE 6.12 Small-signal model for pump clock amplitude regulation.

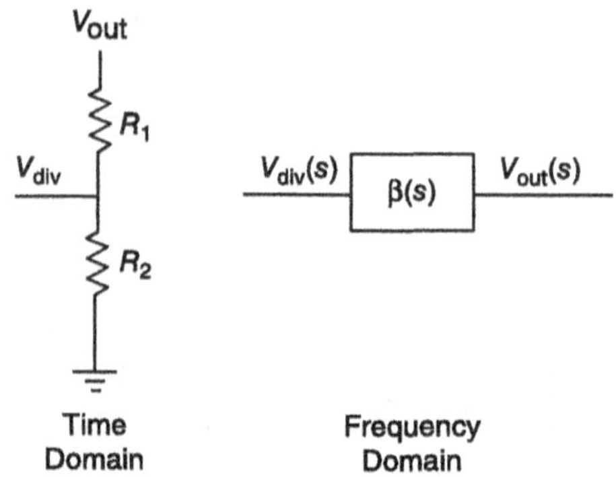

FIGURE 6.13 Feedback ratio β(s).

Mapping components between Figs. 6.10 and 6.12 could help reveal the impact of open-loop gain response $L(s)$ in Eq. (6.39) and the compensation strategy.

$$L(s) = A_{\text{amp}}(s)A_{\text{pump}}(s)\beta(s) \tag{6.39}$$

6.3.2.2 Feedback Ratio

The feedback network could be identified conveniently in both time-domain and frequency-domain representation. Both R_1 and R_2 are assumed to be frequency-independent resistors. From Fig. 6.13, the feedback ratio is given by Eq. (6.40).

$$\frac{V_{\text{div}}}{V_{\text{out}}} = \frac{R_2}{R_1 + R_2} \tag{6.40}$$

In real design, there is parasitic capacitance associated with the physical resistor in design, which implies a pole associated in the feedback network. As the signal travels through the feedback network beyond the given pole frequency, there is attenuation of signal magnitude and phase lagging. If the pole frequencies are within or near the unity-gain frequency ω_t of open-loop gain $L(s)$, it impacts the phase margin of design. Equation (6.41) represents the transfer function of the feedback network in the frequency domain. Beyond resistance and associated capacitance, other components could be added in the feedback network to compensate and improve stability or the dynamic response of the system such as lead compensation.

$$\beta(s) = \frac{V_{\text{div}}(s)}{V_{\text{out}}(s)} = \frac{Z_2(s)}{Z_1(s) + Z_2(s)} \tag{6.41}$$

6.3.2.3 Forward Gain

Figure 6.14 illustrates the forward gain of the system in both the time and frequency domains.

Forward gain $A(s)$ could be expressed as Eq. (6.42) with combination of $A_{amp}(s)$ and $A_{pump}(s)$.

$$A(s) = A_{amp}(s)A_{pump}(s) \tag{6.42}$$

$$A_{amp}(s) = \frac{A_{amp}}{1+\omega/\omega_{amp}} \tag{6.43}$$

The transfer function of the error amplifier could be modeled by Eq. (6.43) with single-pole approximation. Although there are two stages for the error amplifier in Fig. 6.11, it is intuitive to set *NP* as the low frequency pole and keep $V_{clk,\,sup}$ at high frequency with large load current from the clock buffers.

$$A_{pump}(s) = \frac{A_{pump}}{1+\omega/\omega_{pump}} \tag{6.44}$$

$$A_{pump} = K\frac{C}{C_{load}} \tag{6.45}$$

$$\omega_{pump} = 2\pi f_{osc} \tag{6.46}$$

The transfer function for the charge pump is given by Eq. (6.44). For approximation, the dc gain A_{pump} could be expressed as Eq. (6.45), where K is a coefficient to correlate with the number of pump stages, C is the coupling capacitance per stage, and C_{load} is the total capacitance on the output load. The pole frequency of the pump stage is

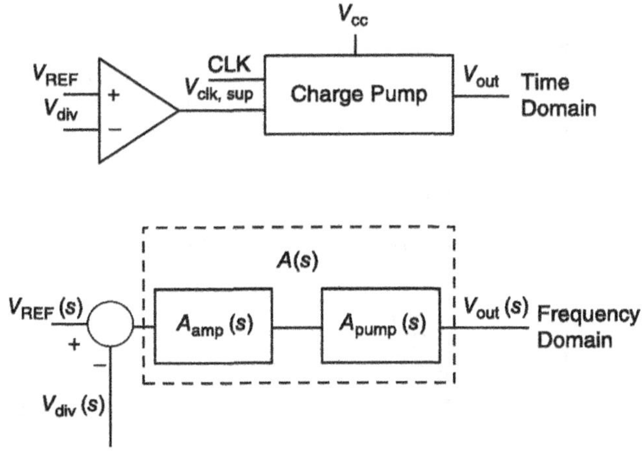

FIGURE 6.14 Forward gain $A(s)$.

related to the pump clock frequency, as shown in Eq. (6.46). It will take at least half a clock cycle or one clock cycle depending on the implementation for the change in the charge pump output to impact the clock amplitude. For simplicity, f_{osc} is used instead.

Forward gain $A(s)$ from Eq. (6.42) can be expanded into Eq. (6.47).

$$A(s) = \frac{A_{amp}}{1+\omega/\omega_{amp}} \frac{KC/C_{load}}{1+\omega/\omega_{pump}} \tag{6.47}$$

6.3.2.4 Stability and Compensation

With known forward gain $A(s)$ and feedback ratio $\beta(s)$, the open-loop gain $L(s)$ can take the form

$$L(s) = \frac{A_{amp}KC/C_{load}[Z_2(s)/Z_2(s)+Z_1(s)]}{(1+\omega/\omega_{amp})(1+\omega/\omega_{pump})} \tag{6.48}$$

Conceptually the analysis of the stability of the charge pump with dynamic voltage scaling is similar to the analysis of any other analog designs. In terms of magnitude, A_{amp} and K contribute to the DC open-loop gain. The feedback ratio of charge pump design is less than 1. It helps in gaining additional phase margin with reduced bandwidth.

In terms of the phase shift, two poles ω_{amp} and ω_{pump} below ω_t in $L(s)$ are given by Eq. (6.48). If those two poles are positioned close to each other, there is potential of 180° phase shift through the loop. Stability is an issue if open-loop gain is still greater than 1.

The stability of the charge pump with dynamic voltage scaling could be analyzed from transient simulations, and compensation techniques presented in early sections can be applied to ensure a stable system with good phase margin.

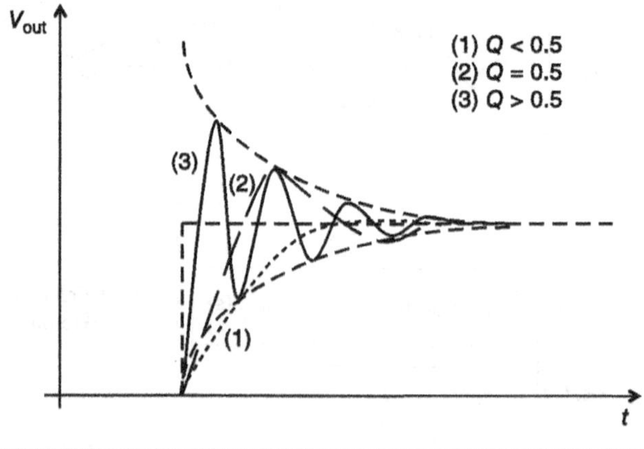

FIGURE 6.15 Step response of charge pump with dynamic voltage scaling on pump clock amplitude.

The step responses of the charge pump with dynamic voltage scaling in the pump clock amplitude are plotted in Fig. 6.15. There are three different step responses for different values of Q. From the dynamic response of the system, case 1 has good phase margin and stability, while case 3 has less phase margin and some ringing in the settling behavior. The step response of system will give insights to the compensation strategy's need to take the knowledge of the locations of poles and zeros.

Phases are impacted by the location of poles and zeros. There are two significant poles located in the loop. One is located in the error amplifier, and the other one is located in the pump stage. With compensation, the separation between ω_{pump} and ω_{amp} should be at least 1 or 2 decades away from each other, and f_{osc} is set to be in the high-frequency range for output capability and area efficiency. The charge pump pole frequency ω_{pump} is fixed by f_{osc} to be in the high-frequency range.

To improve stability, the compensation could be applied to only the error amplifier or feedback network. In order to move ω_{amp} 1 to 2 orders above ω_{pump}, the error amplifier has to consume very high power. This may be prohibitive from the power budget. Technology constraints may not allow design to achieve such a high bandwidth for the error amplifier either. Instead, it would be natural to set ω_{amp} 1 to 2 orders lower than ω_{pump} and make it the dominant pole. Either dominant pole compensation or the Miller-cap compensation technique could be applied.

Another factor impacting stability is open-loop gain. The loop gain is related to the pump stage gain, error amplifier gain, and feedback ratio. The gain in pump stage is fixed by design specification, and it is difficult to change for stability reasons. The gain in error amplifier could be modified. With reduced gain, sometimes it would be easy for loop gain $L(s)$ to reach 0 dB with reduced bandwidth.

The last resort to improve stability could come from the manipulation of the feedback ratio with lead compensation.

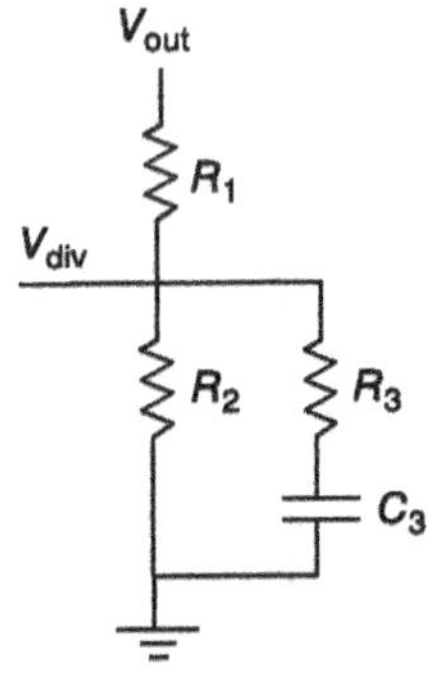

FIGURE 6.16 Compensation in feedback ratio.

Figure 6.16 illustrates an example of compensation in the feedback ratio. Here R_3 and C_3 in series are added in parallel with R_2 in the feedback network. Equation (6.49) illustrates the equivalent impedance $Z_{2eq}(s)$ after compensation.

$$Z_{2eq}(s) = R_2 \frac{1+SC_3R_3}{1+SC_3(R_2+R_3)} \tag{6.49}$$

$$\beta(s) = \frac{R_2}{R_2+R_1} \frac{1+SC_3R_3}{1+SC_3[R_3+(R_2//R_1)]} \tag{6.50}$$

The transfer function for the feedback ratio could be elaborated by Eq. (6.50). The DC gain is identical to the one without compensation. At higher frequency, $\beta(s)$ encounters a lower pole first, then followed by a zero at slightly higher frequency. In open-loop response $L(s)$, the amplitude attenuates at –20 dB/decade after frequency passes $\omega_p = 1/C_3(R_3 + R_2//R_1)$, and it levels off at $\omega_Z = 1/C_3R_3$. Both pole and zero should be positioned within the unity-gain frequency ω_t. Since excess loop gain is attenuated with the addition of a pole/zero pair in the design, a better phase margin is achieved.

$$Z_{1eq}(s) = \frac{R_1}{1+SC_1R_1} \tag{6.51}$$

$$\beta(s) = \frac{R_2}{R_1+R_2} \frac{1+SC_1R_1}{1+SC_1(R_1//R_2)} \tag{6.52}$$

Figure 6.17 illustrates lead compensation in the feedback ratio. Capacitor C_1 is added in parallel with R_1 in the feedback network. The equivalent impedance Z_{1eq} is given by Eq. (6.51), the feedback ratio is modified by the addition of lead compensator C_1, and it is given by Eq. (6.52). The feedback network first encounters a low-frequency zero $\omega_z = 1/C_1R_1$; then it encounters a pole $\omega_p = 1/C_1(R_1//R_2)$ at higher frequency. The added zero should be positioned at slightly higher frequency than unity-gain frequency ω_t to improve the phase margin.

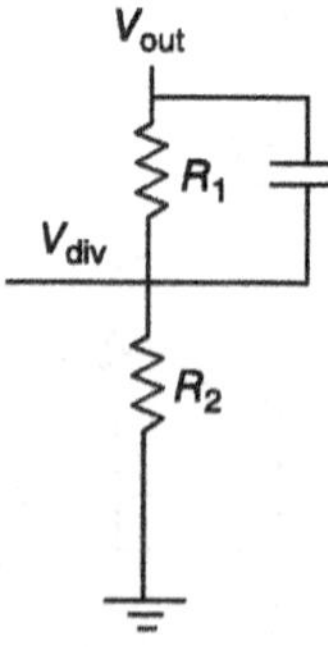

FIGURE 6.17 Lead compensation in feedback ratio.

Information given above provides the basis for compensation of the charge pump regulation. Stability has to be verified by step response in transient simulations. What are some of the rules that can help identify the issues related to stability?

The first observation in waveforms is to examine stage delays in the closed loop. Phase lagging or leading can be mapped with stage delays. With the system being approximated by a second-order system, key emphasis should be on the error amplifier. All the other nodes should have their delays minimized. If the system oscillates or settling takes a longer time, dominant pole frequency should be further lowered, or lead compensation or other techniques should be tried.

The second observation in waveforms is the gain associated with components in the loop. Gain between stages could be measured from the transient waveform. Large gain leads to larger bandwidth, and high-frequency components would start to impact the stability of design. On nondominant pole node, gain could be reduced to enhance the phase margin. For dominant pole node, gain of the stage should not impact stability.

Overall it is possible to analyze the stability of the charge pump with dynamic voltage scaling and to compensate the design with transient simulation only. The optimization is based upon the predefined architecture and common analog compensation techniques. With other charge pump regulations with negative feedback controls, similar techniques could be applied to ensure stability and good phase margin through transient simulations only.

6.4 References

1. Gray, P. R., P. J. Hurst, S. H. Lewis, R. G. Meyer. *Analysis and Design of Analog Integrated Circuits*, 5th ed., John Wiley, Hoboken, NJ, 2009.
2. Carusone, T. C., D. Johns, K. Martin. *Analog Integrated Circuit Design*, 2nd ed., John Wiley, Hoboken, NJ, 2011.
3. Razavi, B. *Design of Analog CMOS Integrated Circuits*, McGraw-Hill, New York, 2000.
4. Allen, P. E., D. R. Holberg. *CMOS Analog Circuit Design*, 2nd ed., Oxford University Press, New York, 2002.

CHAPTER 7

Charge Pump Design, Regulation, and Control by Examples

With the introduction of many design concepts in the preceding chapters, it would be beneficial to show some practical examples to address different aspects of design, such as power efficiency, output noise, and chip size. Depending on the specific design, one of the design metrics would take higher precedence over the others.

7.1 Resistive Divider Feedback Control

Figure 7.1 illustrates the most commonly used feedback network in charge pump design: the resistive divider feedback network, where β is the resistive divider feedback ratio defined in Eq. (7.1). Assuming open-loop gain is much greater than one, output voltage V_{out} can be derived by Eq. (7.2).

$$\beta = \frac{R_2}{R_1 + R_2} \tag{7.1}$$

$$V_{out} = \frac{1}{\beta} V_{ref} \tag{7.2}$$

7.1.1 Accuracy

With only passive element such as resistors included in the feedback network, and with proper layout techniques, the feedback ratio β has little variation over PVT corners. The bandwidth of a pure resistive network is assumed to be high, and ideally it should not impact the open-loop behavior of the design.

As in many analog designs, which rely on the resistor ratio, the base unit resistor size and layout should be uniform in design in order

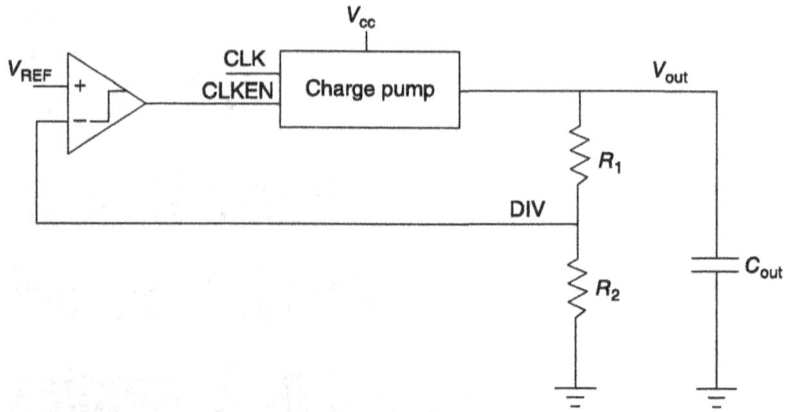

FIGURE 7.1 Charge pump regulation with resistive divider feedback.

to minimize process and lithography variations and mismatches. Resistors with different values from the base unit in design should be derived through either serial or parallel combinations of base unit resistors. This is illustrated in Fig. 7.2. The layout of the unit resistor is repeated in the center with dummy resistors located at both edges.

If a higher degree of matching is required, R_1 and R_2 should be interleaved in layout to minimize local and global mismatches. Dummy resistors at the edges and at the sides of the resistor layout pattern should be included to minimize the edge effect or nonuniform neighboring patterns. This is illustrated in Fig. 7.3. The resistor between A and D and the resistor between B and C are interleaved to minimize the local and global mismatches.

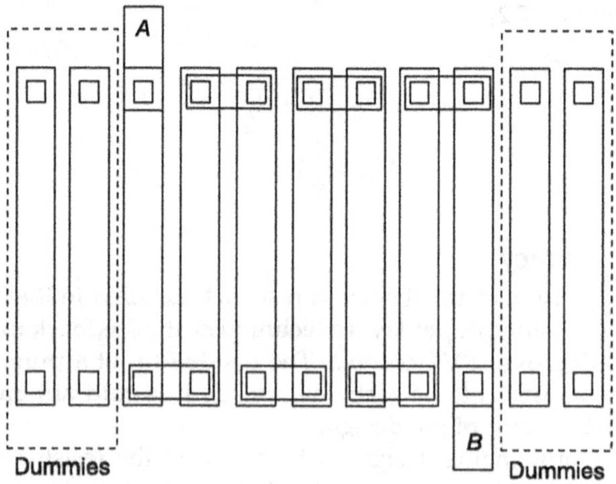

FIGURE 7.2 Resistor layout with unit resistors and dummies.

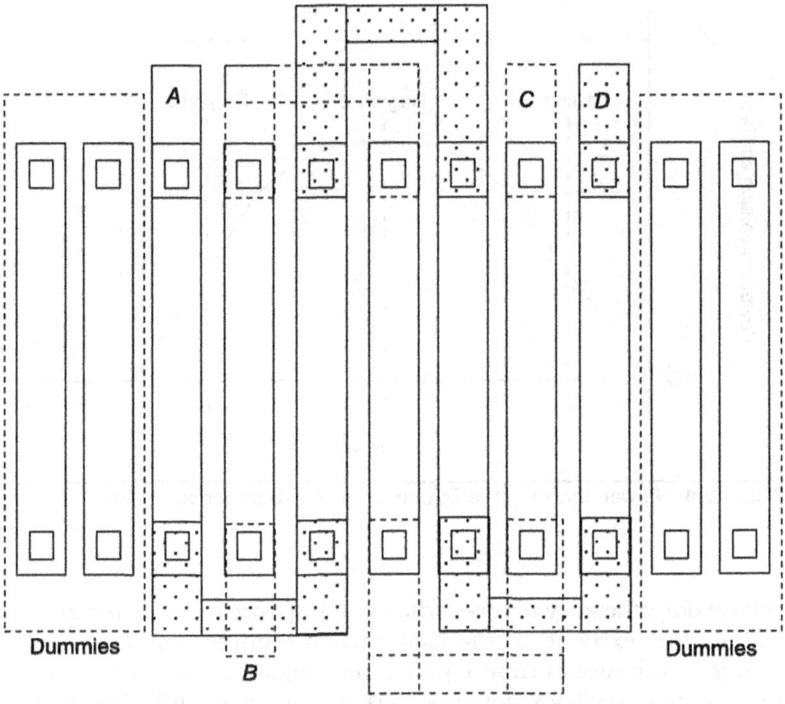

Figure 7.3 Resistor layout with interleaving patterns.

7.1.2 Power

Using the resistor divider feedback does consume quiescent DC power in operations. Equation (7.3) derives DC I_{fb} through the resistor divider. Total DC power consumption P_{fb} is given by Eq. (7.4). As the output voltage V_{out} increases, the power consumption P_{fb} through the feedback network will increase as the square of the output voltage V_{out}.

$$I_{fb} = \frac{V_{out}}{R_1 + R_2} \tag{7.3}$$

$$P_{fb} = \frac{V_{out}^2}{R_1 + R_2} \tag{7.4}$$

Does power consumption P_{fb} matter for the charge pump design? Figure 7.4 illustrates the power conversion efficiency of a DC–DC converter as a function of I_{load}. There are three regions of operation. The power conversion efficiency peaks in region 2. In both regions 1 and 3, the power conversion efficiency drops. With a larger load current such as I_3 in region 3, there would be a large *IR* drop and insufficient charge transfer internal to the charge pump. The power

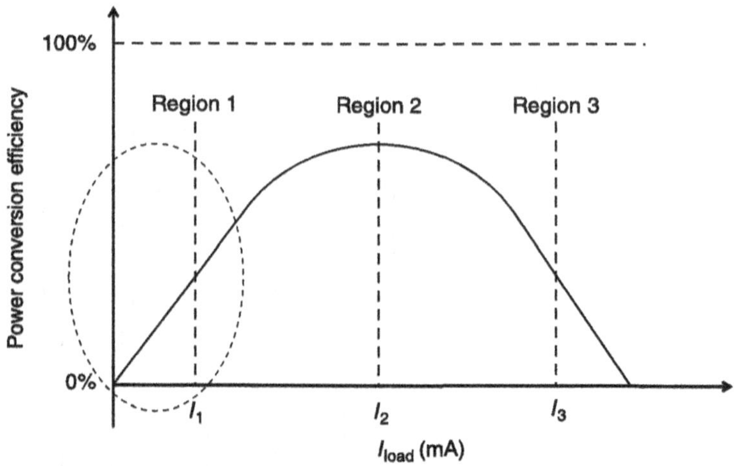

FIGURE 7.4 Power conversion efficiency of a DC-DC power converter.

conversion efficiency reduces when moving from region 2 to region 3. In the other extreme, as the load current reduces from I_2 to I_1, the charge pump moves from region 2 into region 1. In region 1, since the resistive feedback network makes a larger contribution to the quiescent power consumption, the power conversion efficiency reduces as I_{load} reduces.

$$\text{Power conversion efficiency} = \frac{P_{load}}{P_{load} + P_{quiescent} + P_{sw}} \tag{7.5}$$

Equation (7.5) gives the quantitative analysis of the power conversion efficiency for the charge pump. With a resistive feedback network, $P_{quiescent}$ is made up of two components: power consumption of the comparator and power consumption from the feedback network. P_{sw} is the switching power consumption in steady state. As load current decreases, P_{sw} becomes negligible compared with $P_{quiescent}$. Equation (7.5) could be simplified to Eq. (7.6). From Eq. (7.6), as P_{load} decreases with I_{load}, the power conversion efficiency will decrease.

$$\text{Power conversion efficiency} = \frac{P_{load}}{P_{load} + P_{quiescent}} \tag{7.6}$$

For the DC–DC power converter, low-load conditions could account for the majority of operation time. If the power conversion efficiency were low under those conditions, the weighted power conversion efficiency of the system would be low. There is a demand to improve the charge pump power conversion efficiency, particularly at low-load or no-load condition.

7.2 Capacitive Divider Feedback Control

Figure 7.5 illustrates another most commonly used feedback network in charge pump design, which is the capacitive divider feedback, where β in the capacitive divider feedback network is defined in Eq. (7.7). Assuming open-loop gain is much greater than 1, V_{out} can be derived easily from Eq. (7.8). Node DIV is initialized to 0 V before regulation starts in this example.

$$\beta = \frac{1/sC_2}{1/sC_1 + 1/sC_2} = \frac{C_1}{C_1 + C_2} \tag{7.7}$$

$$V_{out} = \frac{1}{\beta} V_{ref} = \frac{C_1 + C_2}{C_1} V_{ref} \tag{7.8}$$

In terms of layout, proper layout techniques are needed to reduce the variation of the capacitive divided ratio from parasitic and process variations. The bandwidth of pure capacitive divider feedback is high, and the feedback network ideally should not alter the phase of the open-loop response of the system.

Similar to other analog circuit designs relying on passive elements, there should be unit capacitance size and layout for the entire design to follow in order to minimize mismatch and variation. Capacitors with different values used in design should be derived from the combination of unit capacitances. This is illustrated in Fig. 7.6 as the base unit is repeated.

If a higher degree of matching is required, C_1 and C_2 should be interleaved in layout to minimize both local and global mismatches and variations. Dummy capacitors at the edges are used to minimize lithography or process variation. The layout with dummy capacitors is illustrated in Fig. 7.7.

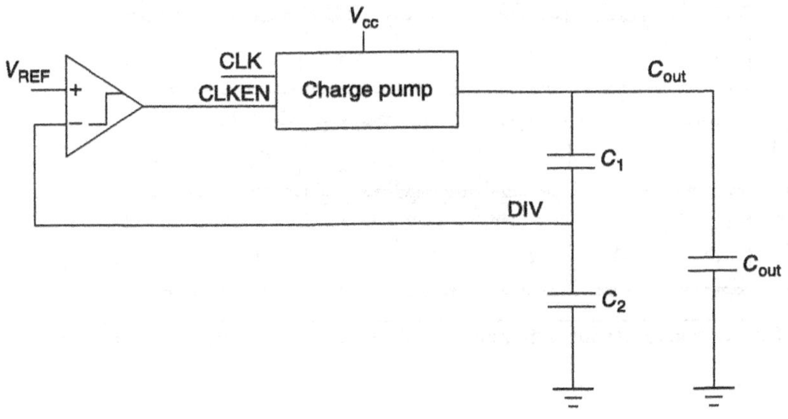

FIGURE 7.5 Charge pump regulation with capacitive divider feedback.

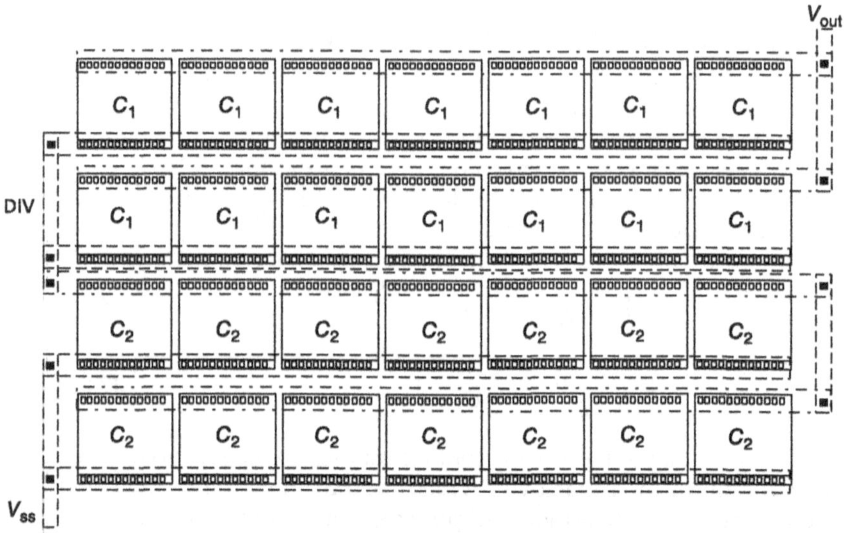

FIGURE 7.6 Capacitor layout with unit capacitors.

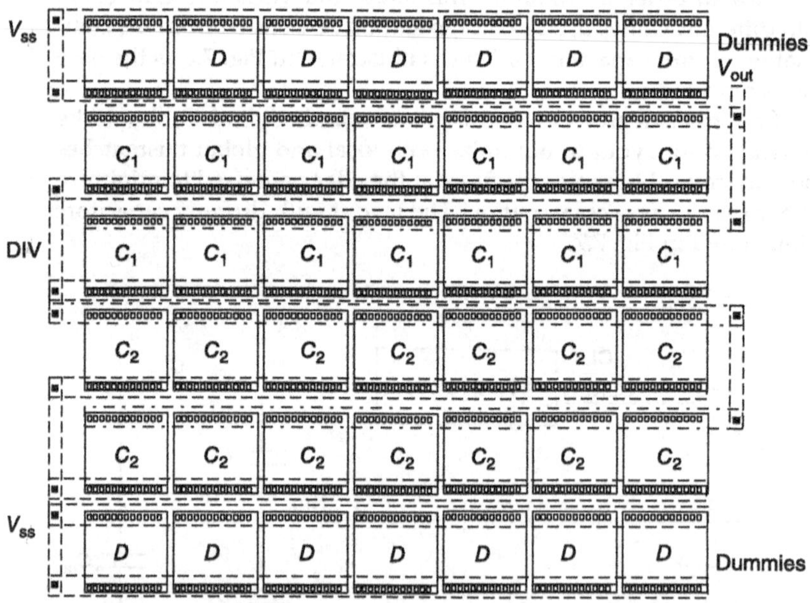

FIGURE 7.7 Capacitor layout with dummies at top and bottom edges of capacitor array.

7.2.1 Accuracy

The accuracy of the capacitance depends on the process variation. In addition, parasitic capacitance always exists from the fringing field, interconnects, and loadings. The accuracy of the capacitor ratio is a function of a process control and layout techniques.

For normal on-chip charge pump design, it is difficult to adopt a good process to have precise control of capacitance. Both systematic variation and random variations exist in designs. The design using a resistor divider feedback network probably has better precision control over the design using the capacitor divider feedback network.

7.2.2 Power

Since the capacitor divider feedback network does not conduct DC as the case exists in the resistor divider feedback network, the design does have a great advantage in minimizing the quiescent power and improve power efficiency. Figure 7.8 illustrates that an overall DC–DC converter with capacitive divider feedback network should have higher power conversion efficiency than one with the resistive divider feedback network in all regions. This is particularly important for charge pump operation with low-load or no-load current condition.

7.2.3 Charge Conservation

One of the concerns about using the capacitive divider pertains to the charge conservation on the divided node over time.

The operation of the capacitive divider feedback is based upon the principle of charge conservation. Since the intermediate node between two capacitors is initialized through some transistors, there

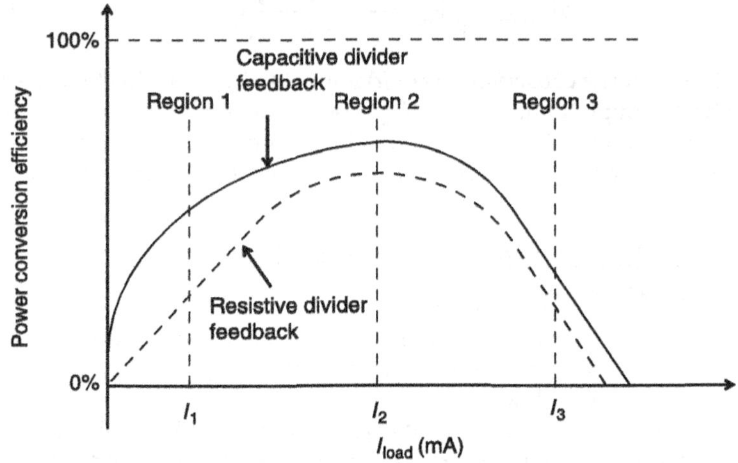

FIGURE 7.8 Power conversion efficiency of DC-DC converter with capacitive divider feedback.

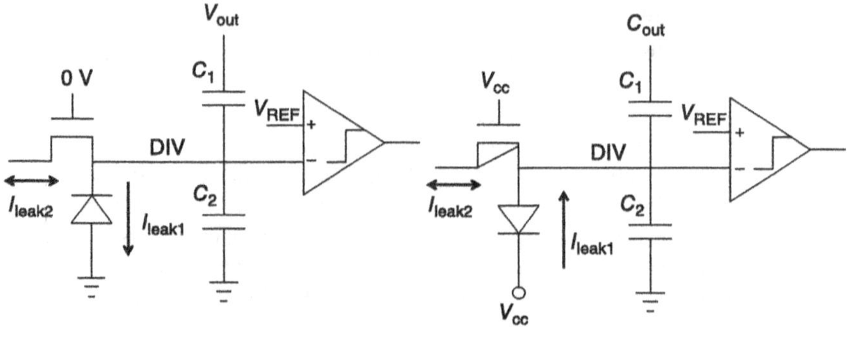

FIGURE 7.9 Leakage current in capacitive divider.

is either an N^+ S/D junction or a P^+ S/D junction connected to node DIV, as illustrated in Fig. 7.9. After initialization transistors being turned off, the charge on node DIV would be held at the moment. Over time, either junction current I_{leak1} or subthreshold current I_{leak2} would leak or inject charge into node DIV. Charge conservation would not hold over time. The error voltage V_{err} over time could be derived by Eq. (7.9). V_{out} would deviate from target due to V_{err}, as illustrated by Eq. (7.10).

$$V_{err} = \frac{(I_{leak1} + I_{leak2})t}{C_1 + C_2 + C_p} \tag{7.9}$$

$$\Delta V_{out,\,err} = \frac{1}{\beta} V_{err} = \frac{C_1 + C_2 + C_p}{C_1} V_{err} \tag{7.10}$$

Since charge conservation could not hold, the operation time has to be short enough to minimize $\Delta V_{out,\,err}$ below one-half LSB in operation. Longer operation should be broken down into shorter sequences with refreshing cycles in between to reset the capacitor divider network.

7.3 Capacitive Divider Feedback without Refreshing Interruption

Using a capacitive divider in the feedback network has the advantage of no DC power consumption from the output supply. This is particularly important for low-power operation.

If operations were long, the junction or subthreshold leakage could inject or leak charge away. Charge conservation would not hold, and error would be introduced in the output. The only option is to interrupt the operations and refresh all the nodes associated with

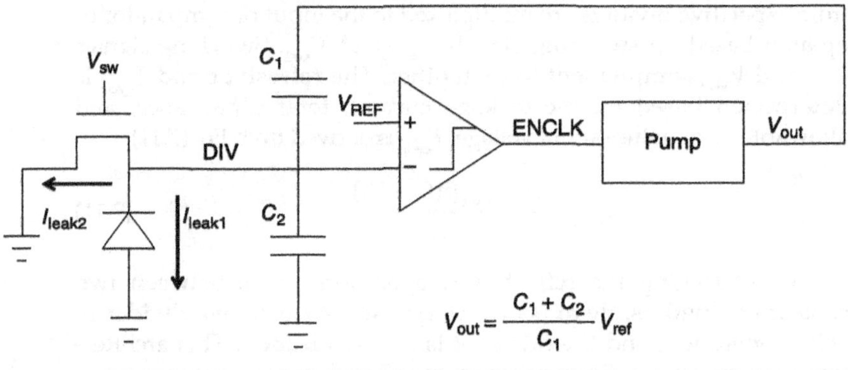

FIGURE 7.10 Typical capacitive divider feedback scheme.

capacitors before the next operation. Interruption means loss of power and performance.

Figure 7.10 illustrates a typical implementation of a capacitive divider feedback scheme with initialization circuits. Figure 7.11 illustrates an implementation of a capacitive divider feedback scheme without the refreshing interruption.[1] Two sets of capacitive dividers are presented for feedback control. When one set of capacitive dividers is operating, the other set is in the resetting state. The feedback voltages

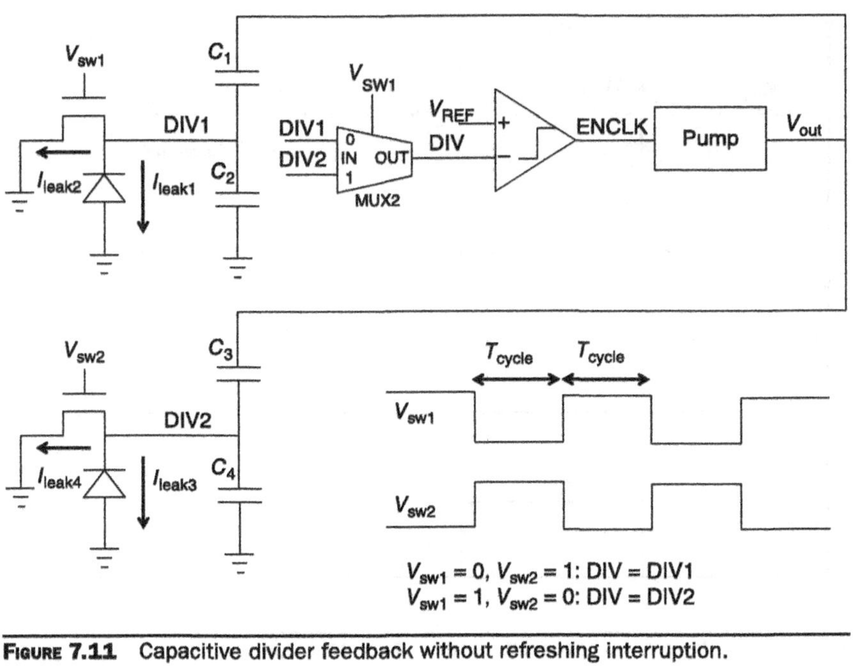

FIGURE 7.11 Capacitive divider feedback without refreshing interruption.

from capacitive dividers are multiplexed to the input of comparator or op-amp based on switching signals V_{sw1} and V_{sw2}. Switching signals V_{sw1} and V_{sw2} complement to each other. The refresh period T_{cycle} is determined based on the leakage current, total capacitance, and allowable error in the output voltage. T_{cycle} is derived from Eq. (7.11).

$$T_{cycle} = \frac{\Delta V_{out,\,err}\,\beta(C_1 + C_2)}{I_{leak}} \tag{7.11}$$

By alternating the refresh and operation cycles between two capacitive dividers, there is always one set of capacitive divider in active operation, and the other set is in initialization. This architecture maximizes the chip performance and improves power saving.

Figure 7.12 illustrates the output waveforms of the given design, where V_{sw1} and V_{sw2} are clocks with opposite clock phases. The duty cycle of the clock is 50%. The duration of T_{cycle} is calculated based upon Eq. (7.11). Assuming leakage would always take charge away from the divided nodes DIV1 or DIV2, closed-loop regulation would force V_{out} to rise in order to compensate the charge loss from leakage on the divided node. At the end of each clock phase, the developed error ΔV_{out} should be within design specification tolerances.

Although this design does take extra area for a duplicate set of capacitive divider and an extra multiplexer, the benefit is improved system performance and reduced power consumption.

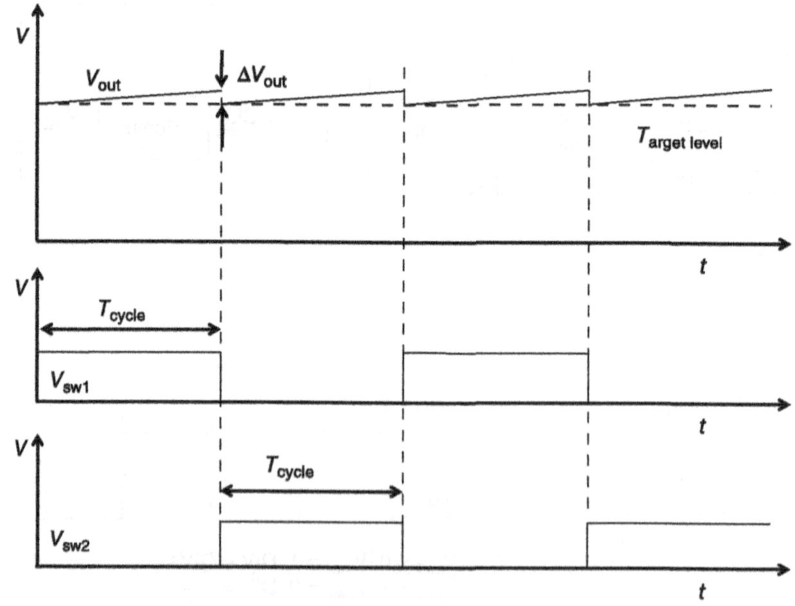

Figure 7.12 Waveforms of regulation without refreshing interruption.

7.4 Floating Regulation

The charge pump may not need to supply current to the load continuously in order to maintain output bias during certain phases in operation. If the operations were known in the design phase, the floating regulation scheme[2] could be implemented to improve the overall power efficiency of the design.

Figure 7.13 illustrates charge pump operations with floating regulation. In normal operation OP, the normal regulation scheme for charge pump is used. In the NO_OP region, the floating regulation scheme is applied. According to Fig. 7.4, if there were no-load current, the power efficiency would drop to nearly 0%. The only way to improve the power efficiency with no-load current is to minimize the quiescent current of the charge pump as much as possible. In the floating regulation scheme, charge pump holds its output bias with load capacitance and decoupling capacitance.

Figure 7.14 illustrates one implementation of the charge pump with floating regulation scheme. Three extra transistors N_1, N_2, and N_3 are added. Transistor N_1 is used to cut off the resistive divider current in the NO_OP region. Transistors N_2 and N_3 are used to isolate high voltage to internal node and to disable the charge pump.

7.4.1 Accuracy

In normal operation, insertion of transistor N_1 has little impact on the feedback ratio if the size is large enough. There is no DC going into the feedback input of the error amplifier. The sizes of N_2 and N_3 can be small to minimize the total parasitic capacitance in the feedback loop in normal operation.

In the NO_OP region, the output of the charge pump is floating. The load capacitance and decoupling capacitance hold output charge. There is junction leakage and subthreshold leakage on the output.

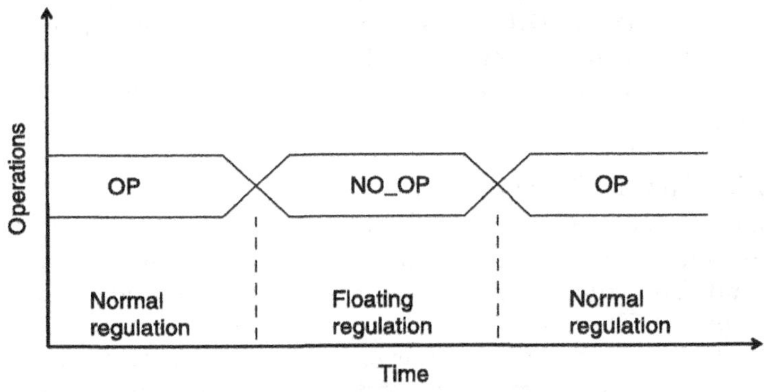

FIGURE 7.13 Charge pump operations with floating regulation.

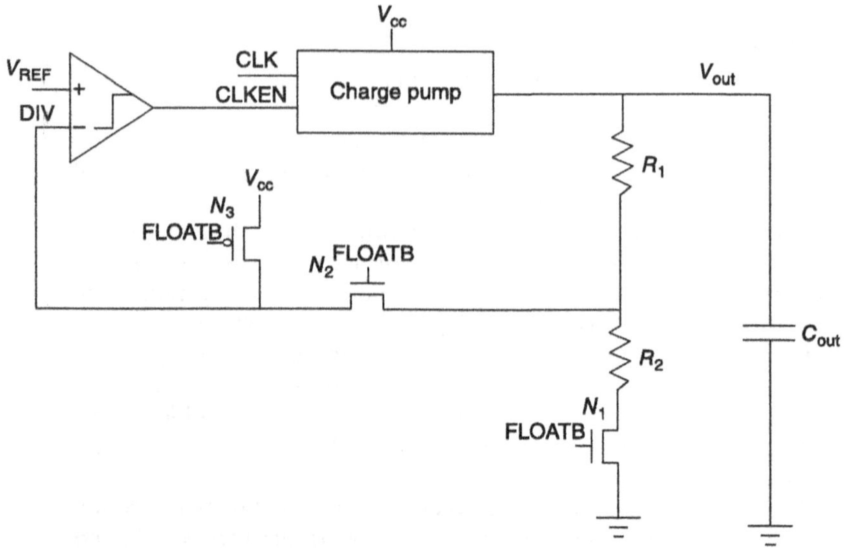

FIGURE 7.14 Charge pump with floating regulation.

Equation (7.12) derives the output voltage error V_{err} over time t; C_{total} is the total capacitance on the output node; and I_{leak} is the summation of all leakage and subthreshold currents on the output.

$$V_{err} = \frac{I_{leak}t}{C_{total}} \tag{7.12}$$

$$T_{refresh} = \frac{V_{err}C_{total}}{I_{leak}} \tag{7.13}$$

Within the NO_OP region, if the error tolerance on the output regulation could be specified, the refresh time period $T_{refresh}$ could be calculated with Eq. (7.13). Operation will resume with the normal regulation scheme to minimize the error term on the output. Figure 7.15 illustrates the sawtooth waveform with floating regulation scheme.

7.4.2 Power Efficiency

In terms of power consumption, since the output is being floated, there is no direct DC path between the charge pump output and ground. The comparator or op-amp could be turned off too to save power. There is only leakage current on the charge pump output. These currents are a few orders lower in magnitude compared with DC in the resistive divider feedback network or biasing circuit.

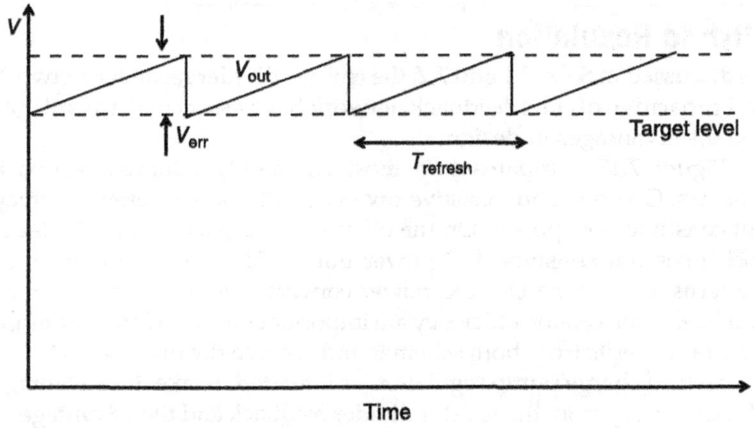

Figure 7.15 Waveform of the charge pump with a floating regulation scheme.

Figure 7.16 illustrates the shifting of the power conversion efficiency curve for the floating regulation scheme at low-load current condition. The charge pump with float regulation scheme defies the normal trend and significantly improves the power conversion efficiency.

$$\% \text{ of Improvement} = \frac{P_{load}/(P_{load} + P_{quiescent2})}{P_{load}/(P_{load} + P_{quiescent1})} = \frac{P_{quiescent1}}{P_{quiescent2}} \approx \infty \tag{7.14}$$

Equation (7.14) illustrates the improvement of power conversion efficiency between the floating scheme and normal scheme. The $P_{quiescent2}$ of the floating scheme is approaching 0 with no DC, while $P_{quiescent1}$ represents the old scheme. The ratio is near infinite.

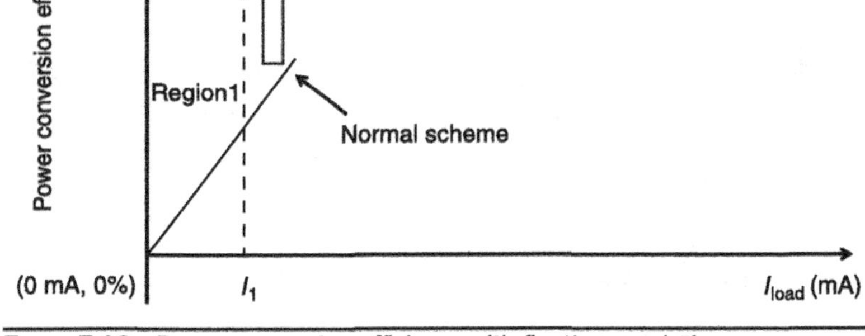

Figure 7.16 Power conversion efficiency with floating regulation.

7.5 Hybrid Regulation

As discussed in Secs. 7.1 and 7.2, the resistor divider feedback network and capacitive divider feedback network have associated advantages and disadvantages in design.

Figure 7.17 compares two most commonly adopted feedback schemes. On one hand, resistive divider feedback has better accuracy but consumes DC power. On the other hand, capacitive divider feedback does not consume DC power but could suffer from accuracy concerns. In modern DC–DC power converter design, both accuracy and power conversion efficiency are important metrics. Can one design adopt the merits from both schemes and remove the drawbacks?

Hybrid charge pump regulation[3] is intended to take the advantage of the accuracy from the resistor divider feedback and the advantage of zero of DC power consumption in regulation from capacitive divider feedback. It combines both the resistive divider feedback network and the capacitive divider feedback network in the control loop. Operations will alternate between two feedback networks to achieve high accuracy and zero DC power consumption.

Figure 7.18 illustrates an implementation of the hybrid regulation architecture. On the right side of Fig. 7.18, both the resistive divider feedback network and the capacitive divider feedback network are presented. In between two divided nodes DIV and DIVR, there is a switch SW that can connect two nodes. The operation of hybrid regulation could be broken into two steps: initialization and regulation.

7.5.1 Initialization Phase

Figure 7.19 illustrates the initialization phase of hybrid regulation. Switch SW is closed in the initialization phase. Nodes DIV and DIVR

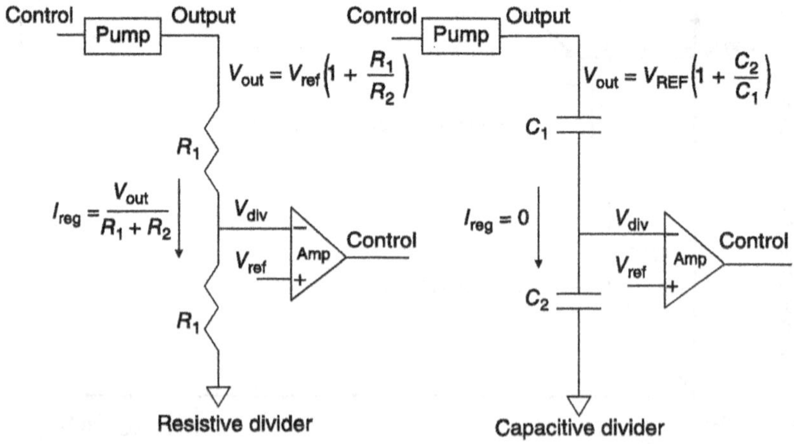

FIGURE 7.17 Resistive divider feedback versus capacitive divider feedback.

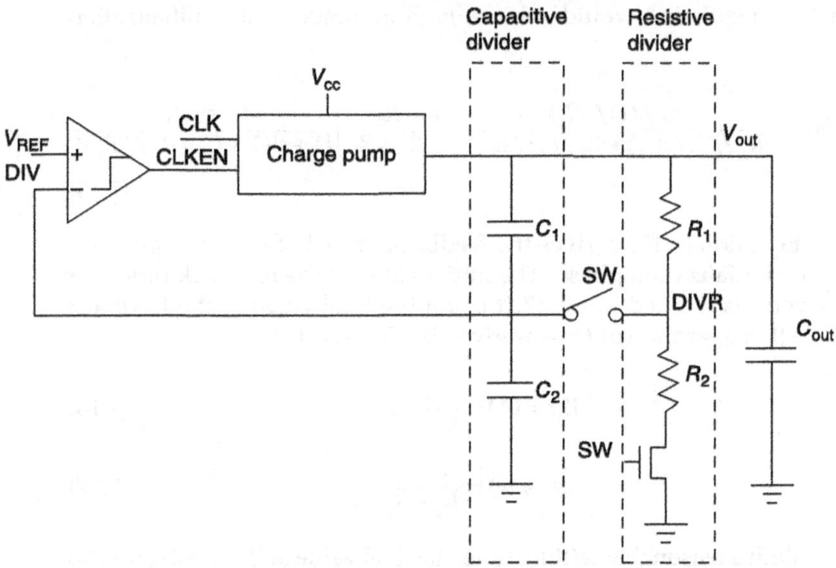

FIGURE 7.18 Hybrid regulation architecture.

are shorted in operation. In terms of feedback control, the capacitive divider influences only start-up and high-frequency operations. The resistive divider feedback network dominates in sampling and feedback control. With the DC accuracy from the resistive divider feedback,

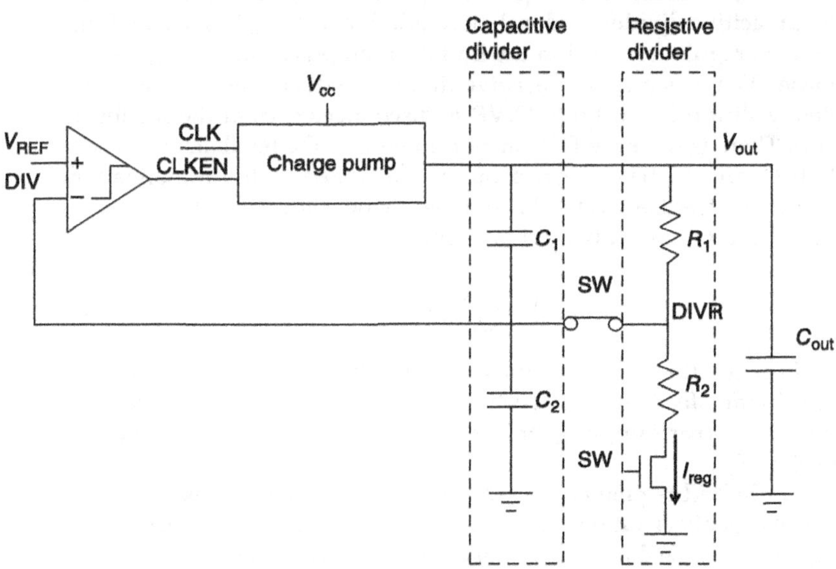

FIGURE 7.19 Initialization of hybrid regulation with the resistive feedback network.

hybrid regulation would achieve high accuracy in the initialization phase.

$$\beta(s) = \frac{R_2 // (1/sC_2)}{R_2 // (1/sC_2) + R_1 // (1/sC_1)} = \left(\frac{R_2}{R_2 + R_1}\right) \frac{1 + sC_1R_1}{[1 + s(C_1 + C_2)(R_1 // R_2)]} \tag{7.15}$$

Equation (7.15) derives the feedback ratio β of hybrid regulation in the initialization phase. The initial value of the feedback ratio β at t = zero$^+$ is defined in Eq. (7.16), and the final value of the feedback ratio β in operation at $t = \infty$ is given by Eq. (7.17).

$$\beta(t = 0^+) = \frac{C_1}{C_1 + C_2} \tag{7.16}$$

$$\beta(t \rightarrow \infty) = \frac{R_2}{R_2 + R_1} \tag{7.17}$$

With a reasonable settling time, the final value of the feedback ratio β is defined by the resistive divider feedback network. The combination of resistive and capacitive divider networks does introduce one zero and one pole in the feedback network in the initialization phase.

Within the initialization phase, V_{out} will settle to the regulation target value base upon resistive divider feedback.

7.5.2 Regulation Phase

Figure 7.20 illustrates the regulation phase of hybrid regulation with the capacitive divider feedback network. Voltage V_{out} has reached the targeted regulation level in the initialization phase. In the regulation phase SW turns off. The resistive divider feedback network is completely shut off, and node DIVR is disconnected from the feedback loop. There is no more DC current flowing in the feedback network. Both C_1 and C_2 have been properly initialized in the initialization phase. The feedback ratio β is defined by only the capacitive divider feedback ratio, as shown in Eq. (7.18).

$$\beta = \frac{C_1}{C_1 + C_2} \tag{7.18}$$

In regulation any error on V_{out} would be sampled by the capacitive divider feedback network. This feedback ratio β is associated with only error sampling, and it is not related to the DC regulation level of V_{out}.

For hybrid regulation, the DC output regulation level is set by the resistive feedback ratio from Eq. (7.17) in the initialization phase. In the regulation phase the feedback ratio impacts only the sampling ratio of the output error. There is additional gain in hybrid regulation design with a higher feedback ratio to minimize the output error or noise.

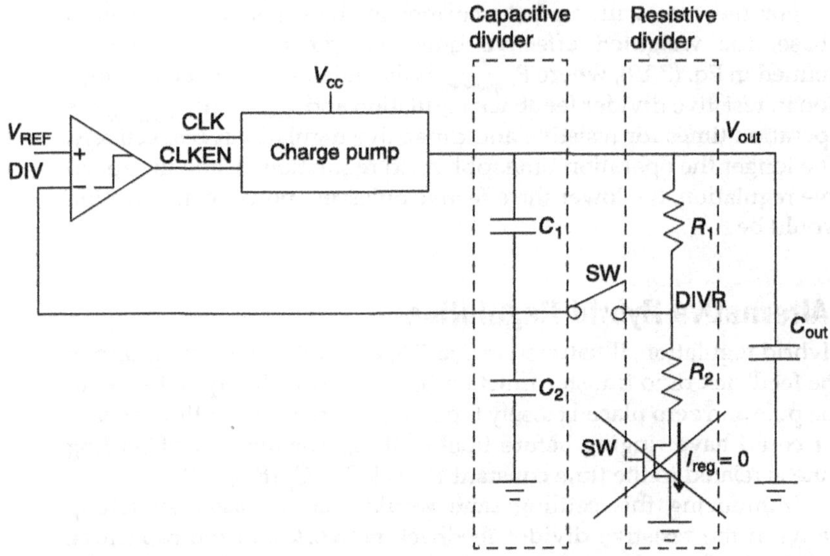

FIGURE 7.20 Regulation phase of hybrid regulation with capacitive divider feedback network.

As shown in Sec. 7.2, the capacitive feedback network cannot hold charge forever on capacitive divided node. Hybrid regulation requires reinitialization (refreshing) based upon error tolerance. By alternating the feedback between the resistive divider feedback network and capacitive divider feedback network, hybrid regulation is realized.

7.5.3 Accuracy

In the initialization phase, the DC accuracy of hybrid regulation is determined only by the resistive divider feedback network with a proper settling time. In the regulation phase, the DC output regulation level is set by the resistive divider feedback network from the initialization phase. The capacitive divider feedback network impacts only the sampling ratio of the output error. It implies the hybrid regulation would be able to achieve the accuracy near that of the resistive divider feedback regulation. In addition, with a higher capacitive sampling ratio, better accuracy is expected with hybrid regulation.

7.5.4 Power

Hybrid regulation alternates between the resistive divider feedback and the capacitive divider feedback. In the initialization phase, the quiescent power is from the resistive divider feedback network. In the regulation phase, there is no DC power consumption in the feedback path.

$$P_{\text{effective}} = \frac{P_{r,\text{quiescent}}\, t_{\text{resistive}}}{t_{\text{resistive}} + t_{\text{capacitive}}} \tag{7.19}$$

For the case with no-load current in the capacitive regulation phase, the weighted effective quiescent power consumption is defined in Eq. (7.19), where $P_{r,\,quiescent}$ is the quiescent power consumption in resistive divider feedback regulation and $t_{resistive}$ and $t_{capacitive}$ are operation times for resistive and capacitive regulations, respectively. The longer the operation time for hybrid regulation to stay in capacitive regulation, the lower the effective quiescent power consumption would be.

7.6 Alternative Hybrid Regulation

Hybrid regulation, illustrated in Sec. 7.5, has both a pole and a zero in the feedback ratio transfer function $\beta(s)$, as shown in Eq. (7.15). With the pole and zero placed closely together, the transient settling behavior could have ringing before final settling. The amount of settling time is related to the time constant $\tau = 1/[(C_1 + C_2)(R_1 \parallel R_2)]$.

Minimizing this settling time would lead to faster switching between the resistive divider feedback network and the capacitive divider feedback network, hence less power consumption. This is the motivation for the implementation of alternative hybrid regulation scheme.[4]

Figure 7.21 illustrates the implementation of the alternative hybrid regulation scheme. In Fig. 7.21, the capacitive divider feedback and the resistive divider feedback networks are separated from each other. Each feedback network has its own comparator or op-amp attached to its feedback-sampling node. A two-input multiplexer to

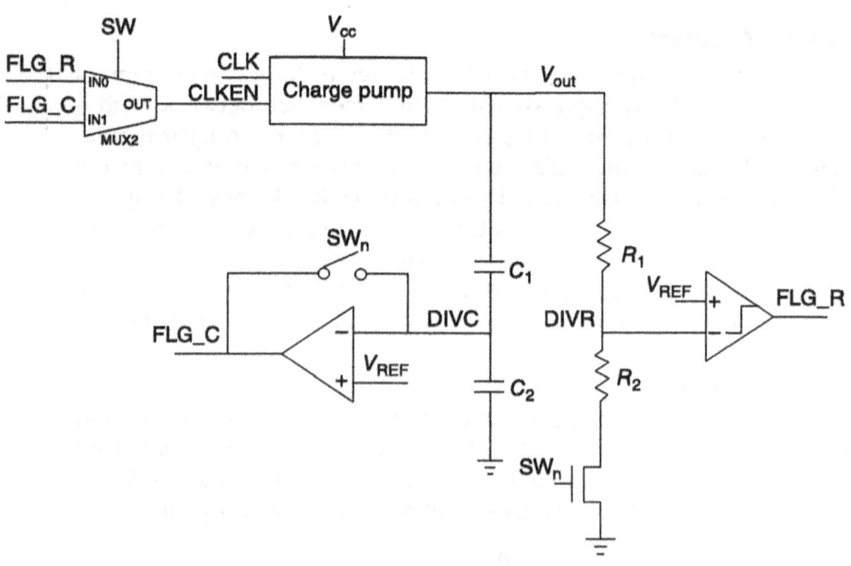

Figure 7.21 Charge pump with alternative hybrid regulation.

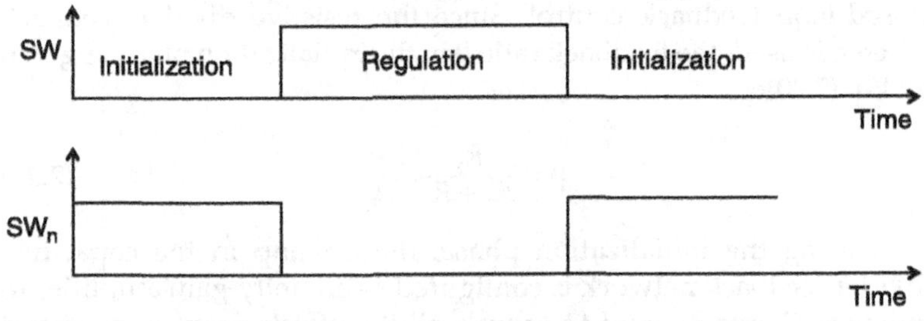

Figure 7.22 Operation of the alternative hybrid regulation.

select which comparator output to be used in feedback operation, either FLG_R or FLG_C.

Figure 7.22 illustrates the operations of the alternative hybrid regulation. The initialization and regulation phases are alternated in operation. In the initialization phase, the resistive divider feedback is used for regulation. In the regulation phase, the capacitive divider feedback is used to minimize DC power consumption. It is easy to illustrate the operations of the alternative hybrid regulation step by step.

7.6.1 Initialization Phase

Figure 7.23 illustrates the initialization phase of the alternative hybrid regulation. The resistive divider feedback network is enabled with SW = 0 and SW_n = 1. The two-input multiplexer selects FLG_R in

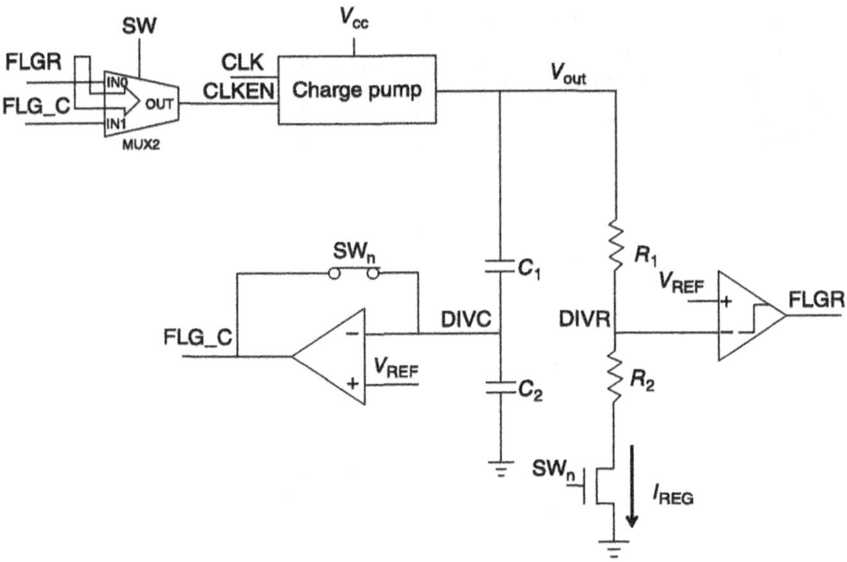

Figure 7.23 Initialization phase of alternative hybrid regulation.

closed-loop feedback control. Since the resistive divider feedback scheme is used, the feedback ratio β in the initialization phase is given by Eq. (7.20):

$$\beta = \frac{R_2}{R_2 + R_1} \tag{7.20}$$

During the initialization phase, the op-amp in the capacitive divider feedback network is configured as an unity-gain amplifier to initialize C_1 and C_2, and to sample all the offsets from op-amp and store them on capacitors. FLG_C is not used in closed-loop feedback control in the initialization phase.

7.6.2 Regulation Phase

Figure 7.24 illustrates the regulation phase of alternative hybrid regulation. With SW = 1 and $SW_n = 0$, the two-input multiplexer chooses FLG_C to close the feedback loop. Switch SW_n is opened, and the unity-gain amplifier is reconfigured into a high-gain comparator. Any sampled error term from V_{out} to DIVC would trigger switching of FLG_C to enable or disable the charge pump. For the resistive divider feedback network, $SW_n = 0$ cuts off the DC path in the resistive divider feedback network. Figure 7.24 does not show the disabling of the comparator associated with the resistive divider feedback network in the regulation phase.

In the regulation phase, Eq. (7.21) derives the feedback ratio β for output error sampling. Any variation in V_{out} would be sampled and

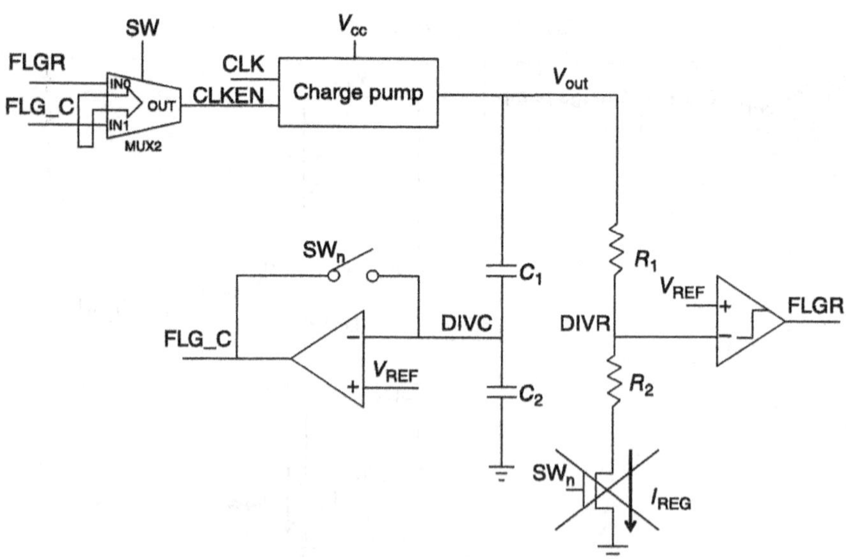

FIGURE 7.24 Regulation phase of alternative hybrid regulation.

be compared with V_{REF} to enable or disable the charge pump. The regulation level of the charge pump output does not depend on β in Eq. (7.21).

$$\beta = \frac{C_1}{C_1 + C_2} \tag{7.21}$$

7.6.3 Advantages

All the benefits from the hybrid regulation architecture highlighted in Sec. 7.5 can apply to the alternative hybrid regulation architecture. There is one additional benefit associated with the new architecture in terms of the settling speed. Equation (7.20) illustrates the feedback ratio β in the initialization phase. It has purely resistive elements. Without additional pole or zero is introduced in the feedback network, the overall settling speed of initialization would be much faster.

The alternative hybrid regulation architecture has the benefits of accuracy, speed, and power.

7.7 Dynamic Frequency Scaling

In certain applications, minimizing the output noise of charge pump would have higher priority over other design metrics. One of the design techniques is to dynamically scale the frequency of charge pump to adapt to the variation of load.

There could two types of dynamic frequency scaling: first is to scale pump frequency base upon known operations; second is to scale frequency dynamically with feedback control.

7.7.1 Dynamic Frequency Scaling with Predefined Operations

Figure 7.25 illustrates dynamic frequency scaling with predefined operations. The charge pump experiences the change of load over predetermined operations, such as initialization, ramp, regulation, and recovery. The required load current in each operation is well defined. Since the output power of the charge pump is proportional to the pump clock frequency, the frequency of the charge pump could be scaled to accommodate various operations. Current I_{out} varies linearly with the pump clock frequency. If I_{out} could be scaled to match I_{load} in each operation, the output noise of the charge pump would be minimized, as shown in Eq. (7.22), where k is a constant much smaller than 1.

$$\Delta V = \frac{(I_{out} - I_{load})t_{cycle}}{C} = k\frac{I_{load}t_{cycle}}{C} \qquad (k \ll 1) \tag{7.22}$$

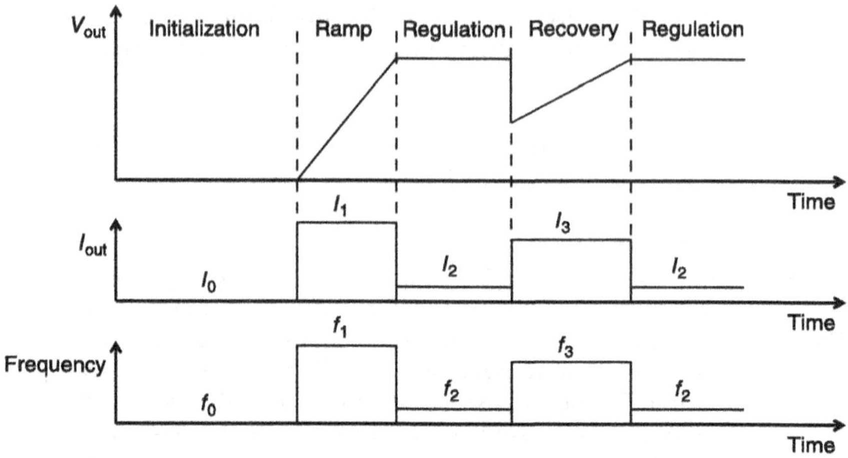

FIGURE 7.25 Dynamic frequency scaling with predefined operations.

7.7.2 Dynamic Frequency Scaling with Feedback Control

Figure 7.26 illustrates the basic scheme of dynamic frequency scaling with feedback control in charge pump operation. The output voltage V_{out} is sampled with the resistive divider feedback network. The voltage difference between node DIVR and V_{REF} is being amplified into

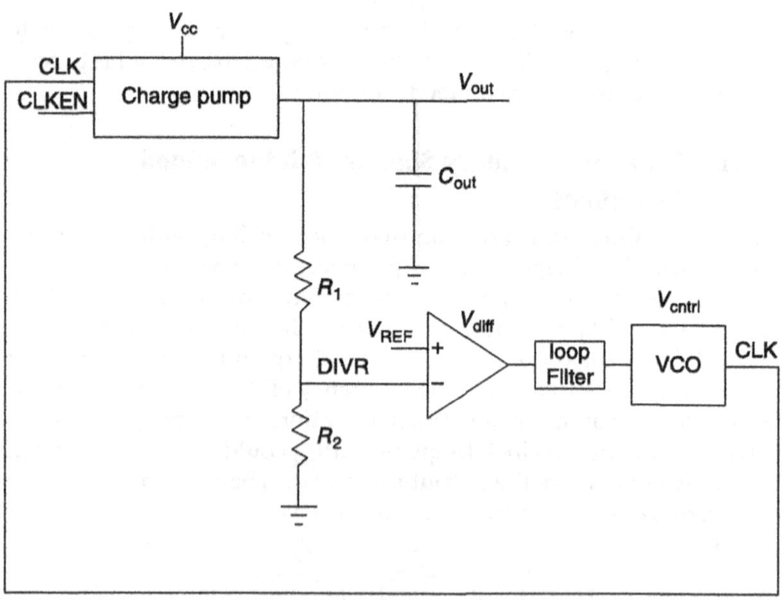

FIGURE 7.26 Charge pump dynamic frequency scaling with feedback control.

V_{diff} which goes into loop filter and generates V_{cntrl}, and it is fed into VCO. The frequency of the pump clock would vary with V_{ctrl} to maintain regulation.

Since the design has negative feedback control, the stability of the charge pump system must be analyzed. Although AC analysis could not be done directly, stability analysis and compensation could still be applied. In the open-loop analysis, only two poles should exist within the unity-gain bandwidth. The first pole should be far below the unity-gain bandwidth, and the second pole should be either at unity-gain bandwidth or at the frequency above. Those two poles must to be separated enough to allow good phase margin and settling behavior. There could be additional LHZ in the control loop to improve stability.

As illustrated in Fig. 7.26, those two poles should be located in the loop filter block and in the charge pump block. All other nodes should have poles located at much higher frequency than the unity-gain bandwidth. In terms of gain, there would be gains associated with the charge pump, error amplifier, loop filter and VCO. Loop gain and the compensation are checked and adjusted through transient simulations.

7.7.2.1 Dynamic Frequency Scaling with Continuous Frequency Range

Figure 7.27 illustrates charge pump frequency scaling with continuous frequency range. The frequency of the charge pump can change continuously between f_1 and f_2, and the output current would be scaled continuously between I_{out1} and I_{out2}.

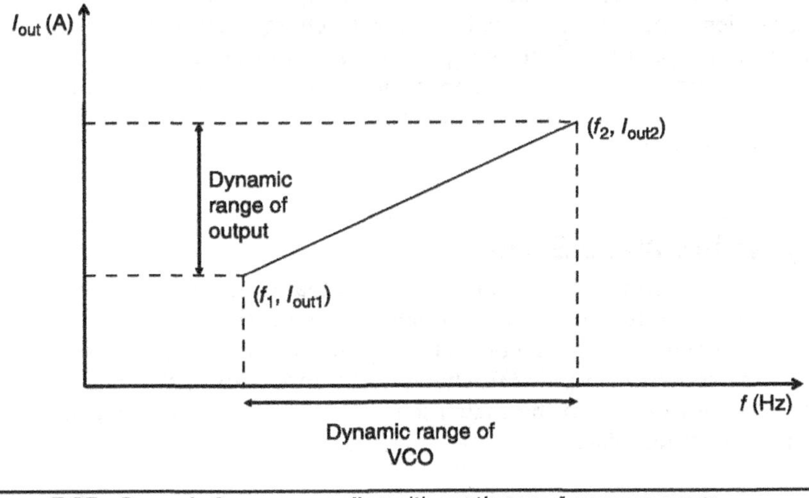

Figure 7.27 Dynamic frequency scaling with continuous frequency range.

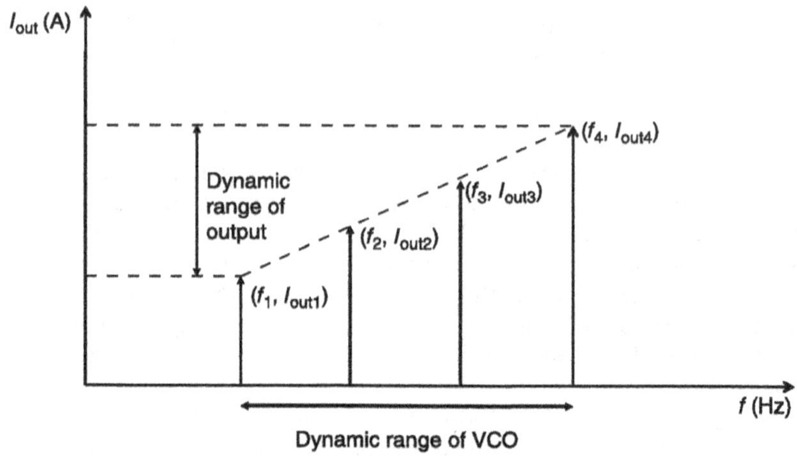

FIGURE 7.28 Dynamic frequency scaling with quantization.

The charge pump output load current can vary by a few orders within operation. If dynamic frequency scaling is used with continuous frequency range, the drawback could be the searching time to lock in the optimum frequency. Overshoot of the output voltage should be a concern in the locking process for high-voltage design.

7.7.2.2 Dynamic Frequency Scaling with Quantized Frequencies

Figure 7.28 illustrates dynamic frequency scaling with quantized frequencies. Instead of searching for the optimum frequency continuously throughout the dynamic range, as in Fig. 7.27, only four fixed frequencies f_1 through f_4 are available for the charge pump to choose. The output capability of the charge pump is quantized into four output ranges from I_{out1} to I_{out4}. Quantization of the frequency range and the output current capability allows faster locking speed and settling behavior for design.

7.8 Dynamic Voltage Scaling

Dynamic voltage scaling is another method that could minimize the charge pump output noise with variation of load. The charge pump output current I_{out} is a function of the pump clock supply voltage $V_{clk,sup}$, as shown in Eq. (7.23). Dynamically modulating the charge pump output power to the match load requirement would help to suppress output noise.

$$I_{out} \propto CV_{clk,\,sup} f_{osc} \tag{7.23}$$

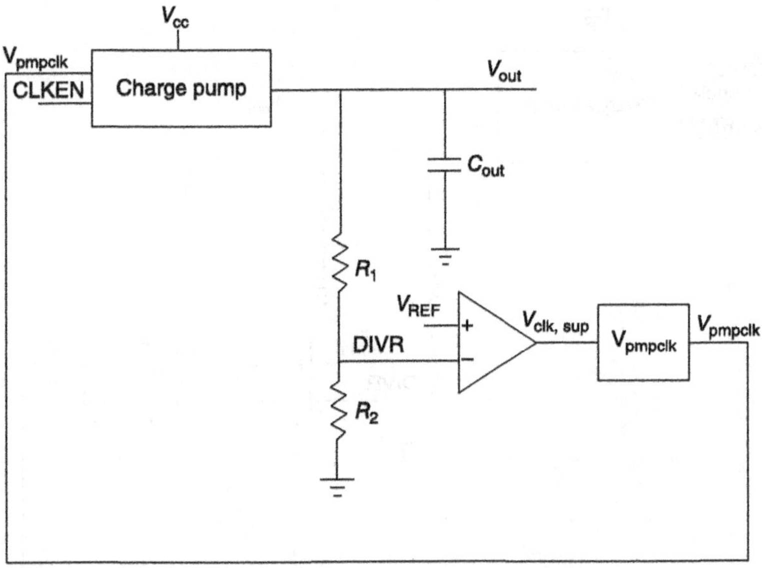

FIGURE 7.29 Charge pump dynamic voltage scaling.

Figure 7.29 illustrates one implementation of charge pump with dynamic voltage scaling. Output voltage V_{out} is being sampled, and the divided voltage DIVR is fed back to the op-amp to control power supply $V_{clk,sup}$ for pump clock drivers. Through the negative feedback loop, the modulated charge pump clock amplitude in turn regulates the output voltage level to target.

In dynamic voltage scaling, the charge pump clock is enabled throughout the entire operation. Power consumption could be a drawback for the design. Figure 7.30 illustrates the implementation of the alternative charge pump dynamic voltage scaling. Instead of modulating the power supply for charge pump clock drivers, the input supply voltage to the first stage of charge pump is modulated. As discussed in Chap. 1, the output current of the charge pump is a function of charge transfer per pump clock cycle. Modulation of charge Q_{in} to the first-stage charge pump through the input supply $V_{sup,in}$ would effectively control the output current I_{out}.

$$I_{out} \propto Q_{in} = CV_{sup,\,in} \tag{7.24}$$

Between the two implementations in Figs. 7.29 and 7.30, there is a difference in power conversion efficiency.

$$P_{quiescent} = C_{clk} V_{clk}^2 f_{osc} + P_{DC} \tag{7.25}$$

Equation (7.25) illustrates the quiescent power consumption in the charge pump dynamic voltage scaling.

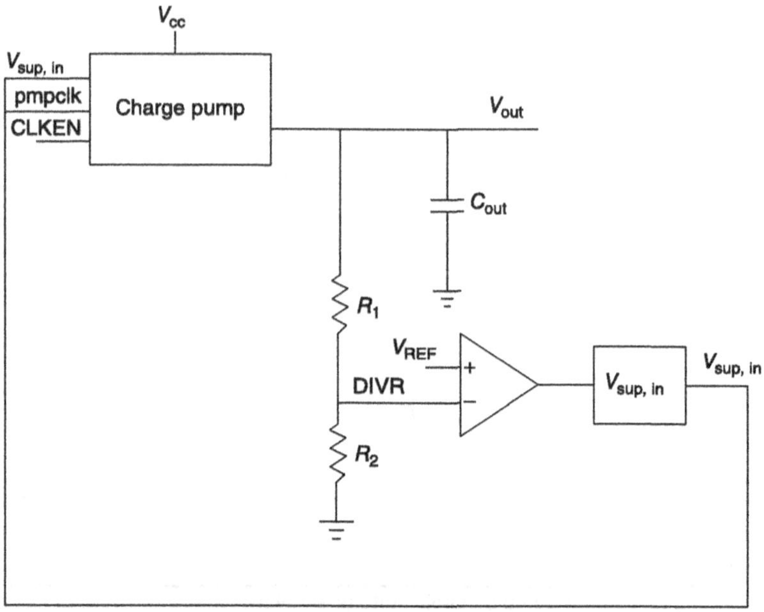

FIGURE 7.30 Alternative charge pump dynamic voltage scaling.

There are two components: One is fixed DC power consumption in the feedback loop and op-amp, and the other is the switching power consumption, which is proportional to the total switching capacitance C_{clk}, switching frequency f_{osc}, and square of pump clock power supply V_{clk}. Since f_{osc} is fixed for dynamic voltage scaling, the switch power consumption has significant weight in total quiescent power consumption.

$$\text{Power conversion efficiency} = \frac{P_{out}}{P_{quienscent} + P_{out}} \approx \frac{P_{out}}{P_{quienscent}} \tag{7.26}$$

With low-load current, the power conversion efficiency could be simplified to Eq. (7.26). It is low because of higher $P_{quiescent}$ power consumption. Between two different implementations, dynamic scaling on the pump clock amplitude would allow a significant reduction in quiescent power.

Overall, dynamic voltage scaling is one of the most effective methods that can minimize charge pump output noise at the expense of power.

7.8.1 Continuous Pump Clock Amplitude Scaling

Figure 7.31 illustrates a charge pump with continuous pump clock amplitude scaling. With the dynamic range of the pump clock amplitude, V_{pmpclk} spans between V_1 and V_2 continuously, and output current I_{out} varies linearly with V_{pmpclk} between I_{out1} and I_{out2}. It is important to

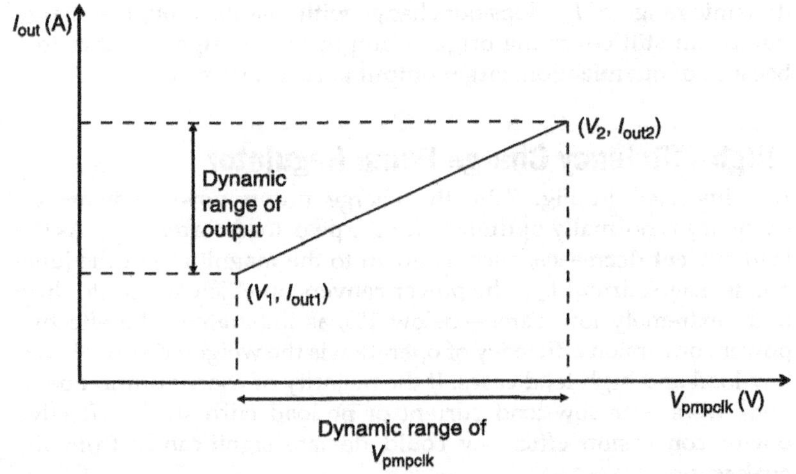

FIGURE 7.31 Charge pump with continuous pump clock amplitude scaling.

design the dynamic range of I_{out} large enough to cover the whole range over which I_{load} could possibly vary in operations.

7.8.2 Quantized Pump Clock Amplitude Scaling

Figure 7.32 illustrates a charge pump with quantized pump clock amplitude scaling. Instead of using the continuous voltage range between V_1 and V_2, only four fixed clock amplitudes are available. The output capability of the charge pump would be quantized into four output ranges between I_{out1} and I_{out4} in this illustration. Since the

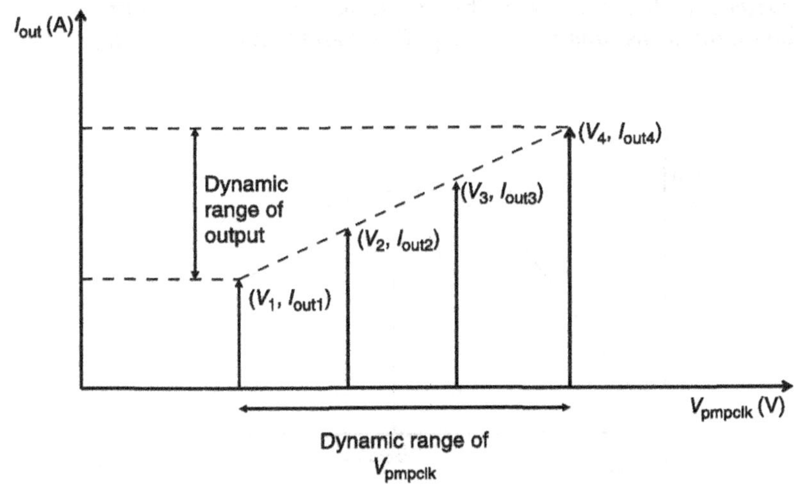

FIGURE 7.32 Charge pump with quantized pump clock amplitude scaling.

dynamic range of I_{out} does not change with quantization, the charge pump can still cover the original output load design specification. Because of quantization, larger output noise is expected.

7.9 High-Efficiency Charge Pump Regulator

As illustrated in Fig. 7.33, the charge pump power conversion efficiency is normally optimized for the peak load current I_{peak}. As the load current decreases, such as down to the magnitude of the junction leakage current I_{leak}, the power conversion efficiency could drop to an extremely low value—below 1%, as illustrated. The effective power conversion efficiency of operation is the weighted sum of both low-load and high-load cases. If the majority of charge pump operations were with low-load current or no-load current, the effective power conversion efficiency could deviate significantly from the peak value.

How to improve the power conversion efficiency with low-load or no-load condition becomes critical.

Instead of using the charge pump to maintain high-voltage regulation with low-load or no-load current in known operations, a different approach could be taken to significantly improve the power conversion efficiency that is limited by the pump architecture.

Figure 7.34*a* illustrates a generic pump with its loading network. The output of the charge pump is sampled by the feedback network, and the comparator is used to maintain the regulation loop. In the illustration, the capacitive feedback network is used to further reduce quiescent DC power consumption. The high open-loop gain ensures the minimum error in regulated output voltage. In terms of loadings, there are three components: PN junction leakage, AC, and DC load current. In the context of this discussion, several assumptions are being made, as illustrated in Eqs. (7.27) and (7.28). Current I_{leak} in the

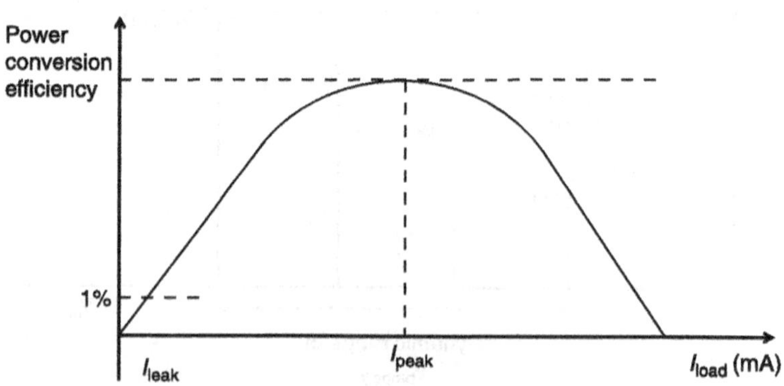

Figure 7.33 Power conversion efficiency of charge pump.

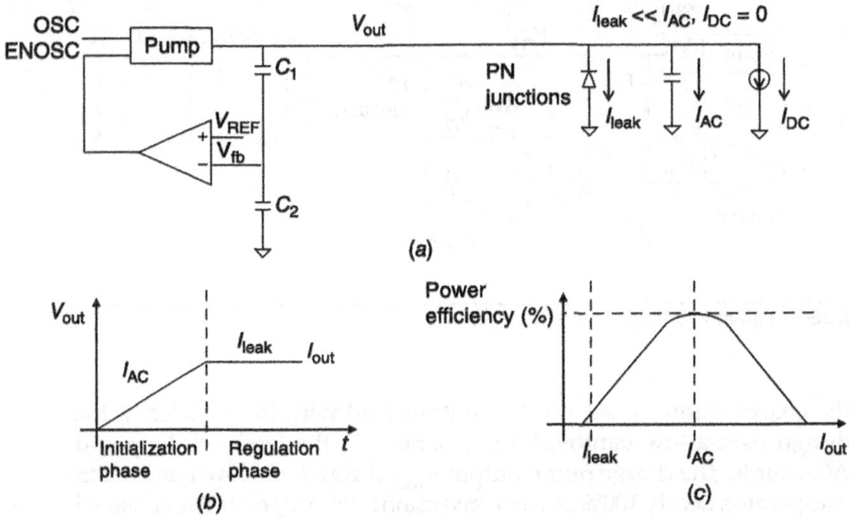

FIGURE 7.34 Power conversion efficiency over different operations.

PN junction diode is assumed to be much smaller than I_{AC}, and I_{DC} is assumed negligible in this analysis.

$$I_{load} = I_{leak} + I_{AC} + I_{DC} \tag{7.27}$$

$$I_{leak} << I_{AC} \qquad I_{DC} = 0 \tag{7.28}$$

Figure 7.34*b* illustrates how the output voltage V_{out} changes over time in operation. Large I_{AC} usually dominates in initialization or load dump recovery phases. After all capacitive loads are charged up, the only current left is the leakage current I_{leak} through reversed-biased PN junction diodes.

Figure 7.34*c* illustrates the familiar power efficiency curve with respect to load current. Since I_{AC} is several orders higher in magnitude than I_{leak}, the circuit design must achieve the highest power conversion efficiency at I_{AC}. After operation switches into the regulation phase, I_{leak} is small, to cause the power efficiency to reduce significantly.

To fundamentally improve the design, the circuit architecture in regulation must be changed. A high-efficiency pump regulator is introduced to address low-power conversion efficiency concerns for design with low-load or no-load current.

Figure 7.35 illustrates the configuration of a high-efficiency pump regulator.[5] In addition to the usual design in Fig. 7.34, two switches SW_1 and SW_2, two capacitors C_3 and C_{AC}, and an additional op-amp2 are added. Switch SW_1 is to connect or disconnect the charge pump from its load. Switch SW_2 is used to configure op-amp2 as a unity-gain amplifier or as a comparator. In the regulation phase, rather than use

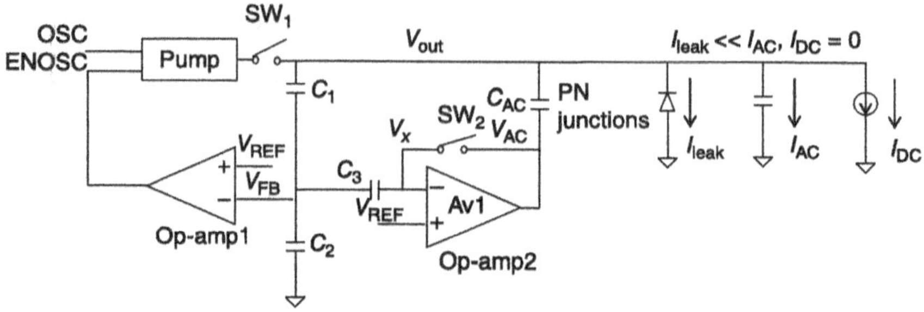

FIGURE 7.35 High-efficiency pump regulator.

the charge pump to supply low output load with low efficiency, the design uses a low bandwidth comparator in the feedback loop and AC-couples the charge pump output V_{out}. Through the low-bandwidth comparator, nearly 100% power conversion efficiency could be achieved in regulation phase.

The analysis of the high-efficiency pump regulator could be done in two steps: the initialization phase and the regulation phase.

7.9.1 Initialization Phase

Figure 7.36 illustrates the initialization phase of the high-efficiency pump regulator. In Fig. 7.36*a*, both switches SW_1 and SW_2 are closed. The charge pump uses the original feedback path to regulate and

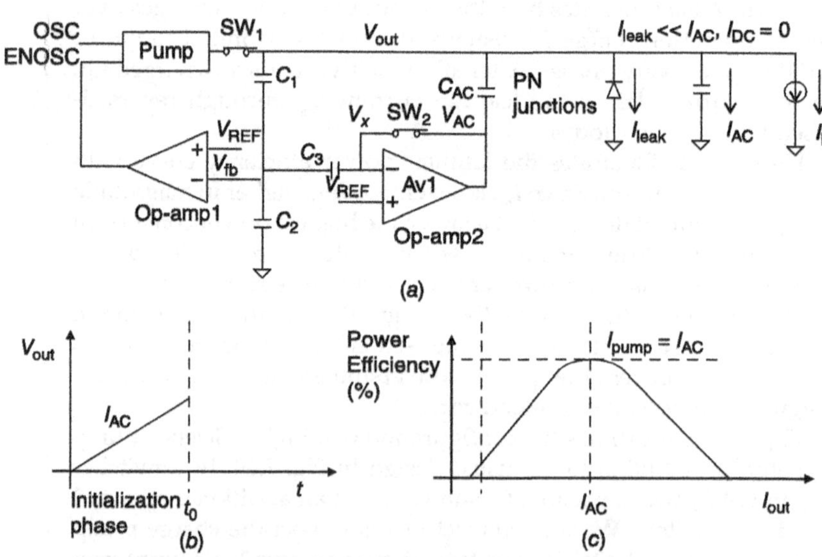

FIGURE 7.36 Initialization phase of high-efficiency pump regulator.

supply I_{AC}, as shown in Fig. 7.36*b*. Under this condition the charge pump achieves peak power conversion efficiency, as shown in Fig. 7.36*c*. Op-amp2 is configured as a unity-gain amplifier. Once output reaches regulation, voltages across C_3 and C_{AC} are initialized properly. Any offset errors and mismatch would be stored on C_3 at end of the initialization phase.

7.9.2 Regulation Phase

Figure 7.37 illustrates the regulation phase of the high-efficiency pump regulator. Switches SW_1 and SW_2 are opened. Charge pump and op-amp1 are turned off and disconnected from the load. Op-amp2 is configured as a high-gain comparator. A new negative feedback loop is formed to maintain regulation of V_{out}.

With the new negative feedback loop, any variation of V_{out} would be sampled and corrected by op-amp2 through AC coupling between V_{AC} and V_{out}.

Figure 7.38 illustrates the regulation of the charge pump output with a second op-amp. Current I_{leak} represents the total load current in the regulation phase of the charge pump. Current could be sourcing from V_{out}, or it could be injected into V_{out}. The implementation of op-amp in Fig. 7.38 is a one-stage op-amp with differential input and single-ended output. It is for illustration purposes. With low-load

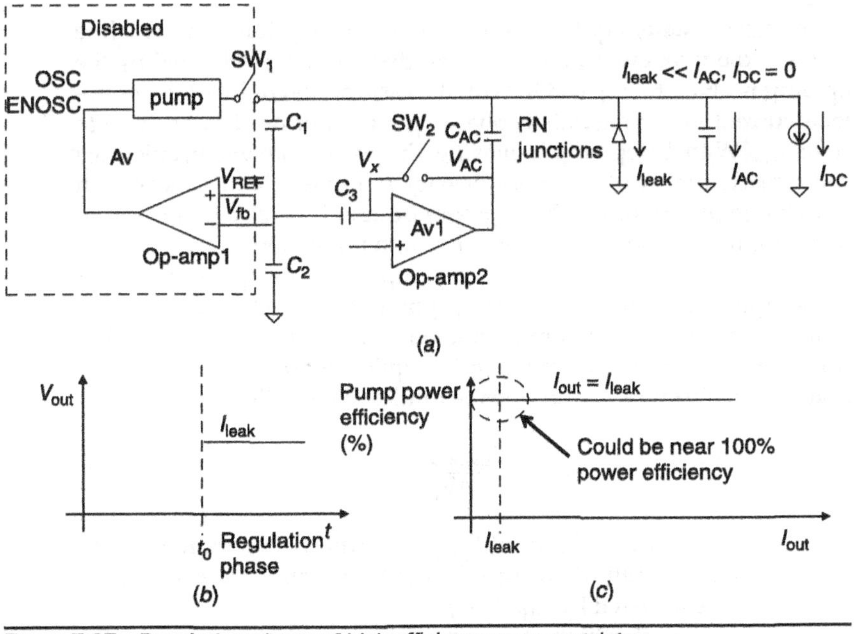

FIGURE 7.37 Regulation phase of high-efficiency pump regulator.

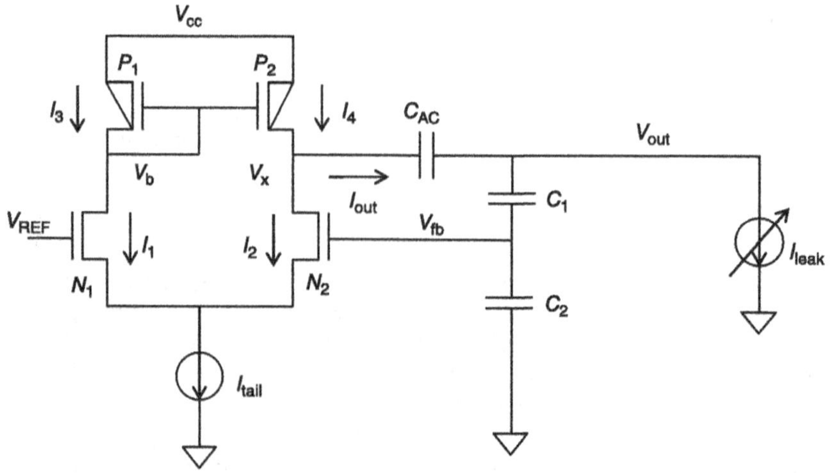

FIGURE 7.38 Regulation of charge pump output with second op-amp.

current in regulation and low-speed operation, the op-amp should have a low-bandwidth design, to minimize power consumption.

$$I_{out} = I_{leak} = I_4 - I_2 \tag{7.29}$$

$$|I_{out,\,max}| = I_{tail} \tag{7.30}$$

In steady state, Eq. (7.29) should hold. At any time, the absolute value of the maximum output current that could be delivered by the op-amp is given by Eq. (7.30). With the assumption of low-load or no-load current in the regulation phase, I_{tail} does not need to be large to cover I_{leak}. With the op-amp in slewing, the power conversion efficiency could reach almost 100%. This efficiency is impossible to achieve with any charge pump design. With proper design, I_{tail} is at the same order of magnitude as the output load current in regulation. Since I_{tail} of the op-amp is sourcing directly from input supply V_{cc}, the quiescent power consumption is small. For the charge pump, the current sourcing from input supply V_{cc} to maintain regulation could be a few orders higher from the load current in regulation. The quiescent power consumption is large. Architectures determine the fundamental difference.

$$C_{AC} = \frac{I_{leak,\,max}\Delta t}{\Delta V_x} \tag{7.31}$$

With maximum load current $I_{leak,\,max}$, duration of operation Δt, and dynamic range of output swing ΔV_x from the op-amp, the capacitance C_{AC} could be estimated by Eq. (7.31).

In summary, the high-efficiency pump regulator improves the power conversion efficiency of design with low-load or no-load current in regulation. The added low-bandwidth op-amp, capacitors, and switches have minimum area impact on the given design.

7.10 RC-Based V_T Cancellation

There are many types of V_T cancellation architectures in practice. Despite design complexities, normally a reasonable good V_T cancellation scheme would consume about 5% to 10% of the total power consumption. Can the power consumption be further reduced?

In terms of V_T cancellation, two major factors contribute to the additional quiescent power consumption. The first factor is related to the efficiency in initialization of the gates of passing elements. For NMOS, the higher the initialization voltage, the lesser power it took for V_T cancellation. The second factor is related to the parasitic capacitance associated with the V_T cancellation scheme. Coupling efficiency is degraded with parasitic capacitance. Switching power consumption is a linear function of total parasitic capacitance.

The *RC*-based V_T cancellation scheme is intended to achieve the higher initialization level with minimum effort and to minimize the parasitic capacitance associated with coupling and switching.

The operating principle of the *RC*-based V_T cancellation scheme[6] is the combination of the low-pass function and high-pass function. The low-pass function establishes the DC average, and the high-pass function allows AC coupling for V_T cancellation.

7.10.1 Low-Pass Function

Figure 7.39 illustrates a low-pass function of signal transmitted from V_{in} to V_{out}. Voltage V_{in} is a step function that switches between 0 and 1 V at time t_0. Voltage V_{out} has linear settling behavior over time. The

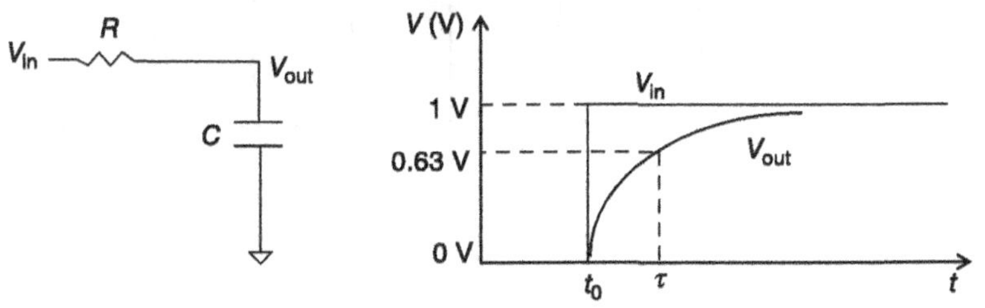

t/τ	V_{out}/V_{final}	Error
0	0%	100%
1	63%	37%
2	86%	14%
3	95%	5%
4	98%	2%
5	99%	1%

FIGURE 7.39 Low-pass function.

table in Fig. 7.39 shows the normalized settling behavior and error term. After 1τ, V_{out} reaches 63% of target. It would take 4τ to reach 98% of final target. The low-frequency component in the input signal would take at least 3τ to 4τ to be passed to output.

7.10.2 High-Pass Function

Figure 7.40 illustrates a high-pass function. Voltage V_{out} has been initialized to 0 V through *R*. Voltage V_{in} is a square pulse function that switches between 0 and 1 V, as shown in Fig. 7.40*a*. Output voltage V_{out} has linear settling behavior, as shown in Fig. 7.40*b*. It switches between the proximities of 0 and 1 V.

If different initialization voltages were applied, the output waveform would shift vertically up or down with respect to different DC voltage, while AC transient behavior would be identical. This phenomenon is illustrated in Fig. 7.41. Two cases are plotted with the initialization voltage of 5 V and 10 V. V_{in} is clocking between 0 V and 1 V with zero transition time. V_{out} in Fig. 7.41*a* switches between 5 V and 6 V respectively, while V_{out} in Fig. 7.41*b* switches between 10 V and 11 V respectively.

7.10.3 Combined Effects

Figure 7.42 illustrates the implementation of the RC-based V_T cancellation scheme. The right side of the block is the normal path. It has conventional diode-connected NMOS N_1 connected between preceding

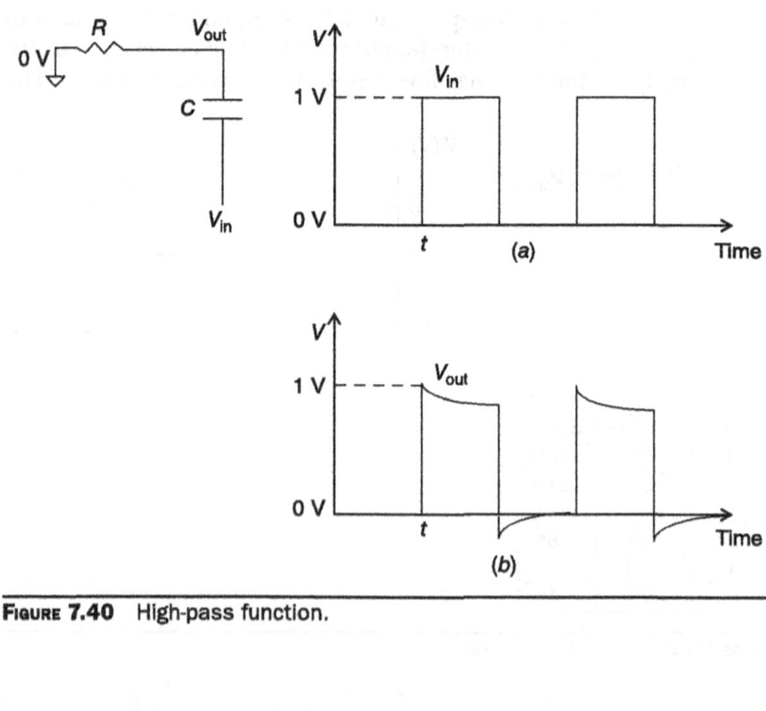

FIGURE 7.40 High-pass function.

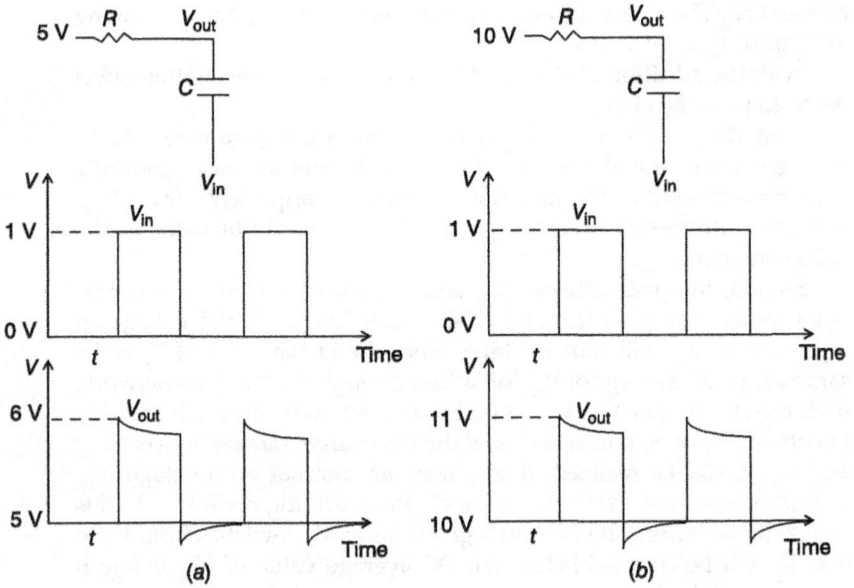

FIGURE 7.41 High pass with different DC initialization.

stage V_{n-1} and output V_{out}. On the left side, three new components are added for V_T cancellation. NMOS N_2 has its drain tied to preceding stage V_{n-1}, and its source node is tied to output V_{out}. Its gate is connected to node V_g, which connects to capacitor C and resistor R. Capacitor C has its top plate connected to node V_g and its bottom plate tied to node V_{clk}. Resistor R is connected between node V_{out} and node V_g.

In operation, without the V_T cancellation circuit block on the left side, V_{out} will rise as V_{n-1} rises through the diode-connected passing

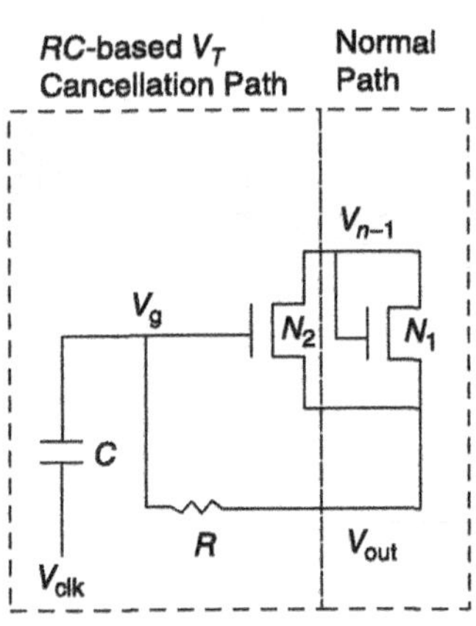

FIGURE 7.42 RC-based V_T cancellation.

element N_1. There is a voltage difference between V_{n-1} and V_{out} owing to V_T and V_{ov} of N_1 and I_{load}.

With the addition of R, C, and V_{clk}, there is a V_T cancellation effect on N_2 to pass the charge.

First, the path between V_{out} and V_g forms a low-pass network. As V_{out} rises over time, the potential on node V_g will increase gradually over time to extract DC and low-frequency components from V_{out}. The lower bound DC average potential of V_g would be same as the DC average of V_{out}.

Second, the path between V_{clk} and V_g forms a high-pass network. After the normal path from Fig. 7.42 establishes the initial voltage on V_{out}, voltage V_{clk} will start clocking. Since the DC average of V_g is the same as the DC average of V_{out} initialized through the low-pass network, with the clocking of V_{clk}, V_g would be coupled up by an additional V_{cc}. Effectively V_T of N_2 is canceled, and the full charge transfer between V_{n-1} and V_{out} could be realized. If V_{n-1} were an internal pump stage, V_{clk} should be synchronized with the clock phase driving node V_{n-1}. In this case, with linear settling of V_g in high phase, as V_{clk} switches from V_{cc} to 0 V, V_g will be coupled below the DC average value of V_{out} before it recovers. It helps to cut off reverse leakage current in operation.

As V_{out} rises with the help of V_T cancellation on N_2, a positive feedback loop is established in Fig. 7.42. Figure 7.43 illustrates

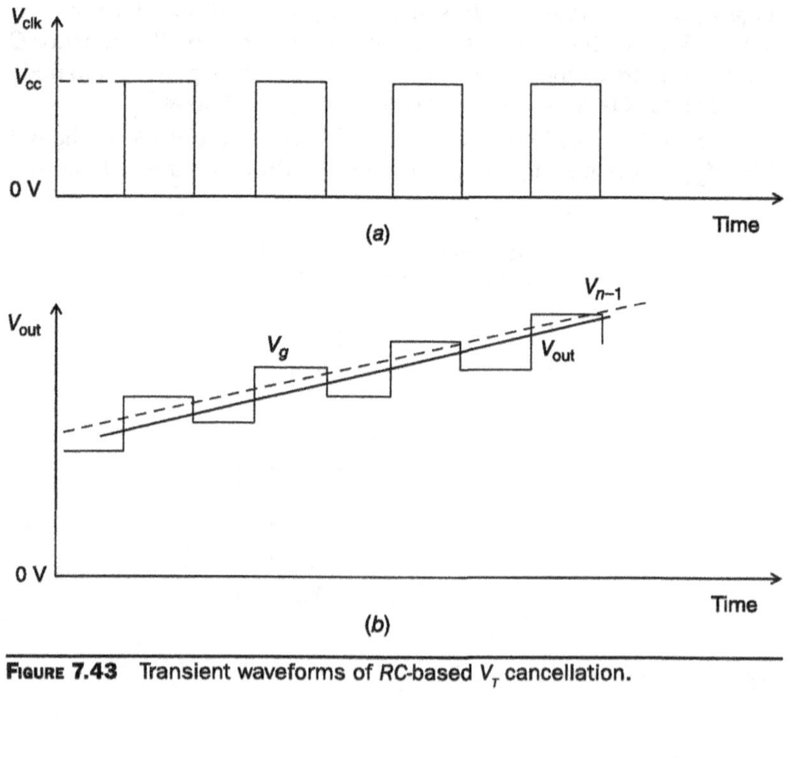

Figure 7.43 Transient waveforms of *RC*-based V_T cancellation.

transient waveforms of RC-based V_T cancellation. Over the course of operation, positive feedback will equalize V_{out} and V_{n-1} at the end. As V_{n-1} rises, both the DC average of V_g and V_{out} will rise.

There are two motivations to use the passive element R only in an RC-based V_T cancellation scheme.

In any known V_T cancellation architectures, the efficiency of initializing the gate of the passing element impacts both performance and power consumption. The higher potential the gate could reach during initialization phase, the lower the effort needed for capacitive coupling. If active devices were used, there would be switching activities and overdrive requirements in order to move charge onto the gate of passing elements. The implication is power consumption and even higher potential is required. In terms of using passive elements R in the loop, there is no switching power and higher potential required to initialize the gate bias.

With higher initialized gate voltage, the dimension of the passing elements like N_1 and N_2 could be reduced with the help of higher g_m. Lower capacitive load leads to better coupling ratio, smaller C, and a clock driver.

All the benefits derived from passive element usage reduce the quiescent power consumption of RC-based V_T cancellation compared with ones using active elements.

7.11 Charge Pump Clock Equalization

To reduce quiescent power consumption in the charge pump, one area of focus would be the switching power of the clock drivers. When output from digital circuits switch between two power supply rails, charge is wasted through charging and discharging activities over cycles. If some portion of charge could be recycled in operations somehow, rather than discharging directly to ground, then switching power consumption could potentially be reduced.

Figure 7.44 illustrates the equivalent pump clock drivers and capacitive loadings. In operations, CLK_1 and CLK_2 charge and discharge C_{EQ1} and C_{EQ2} in alternating clock phases. Figure 7.45 illustrates the complementary pump clock waveforms. While CLK_1 switches from 0 to V_{cc} to charge up its capacitive load, CLK_2 switches from V_{cc} to 0 to discharge and reset its capacitor associated. Rather than dumping all the charge to ground, if some portion of the charge could be transferred from the output of one clock output to the other through the equalization process, then a certain amount of charge could be recycled.

Figure 7.46 illustrates one implementation of pump clock drivers with charge equalization.[7] Compared with the circuit implementation in Fig. 7.44, the clock output drivers were modified to be tristate buffers, and an equalization transistor connects CLK_1 and CLK_2. Without equalization, pump clock drivers operate normally.

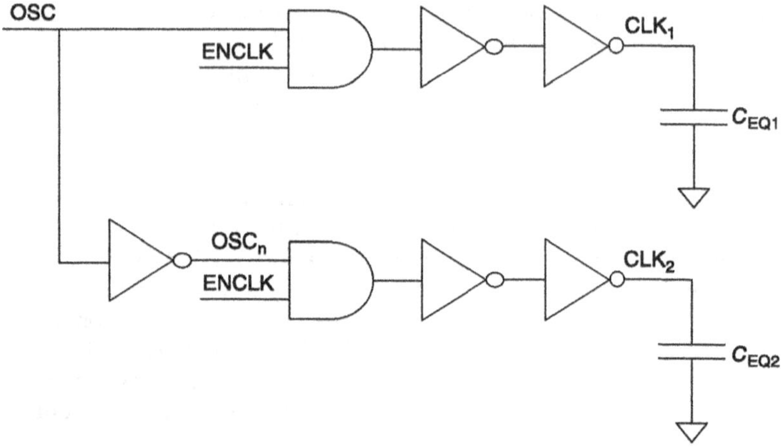

Figure 7.44 Equivalent pump clock drivers and loadings.

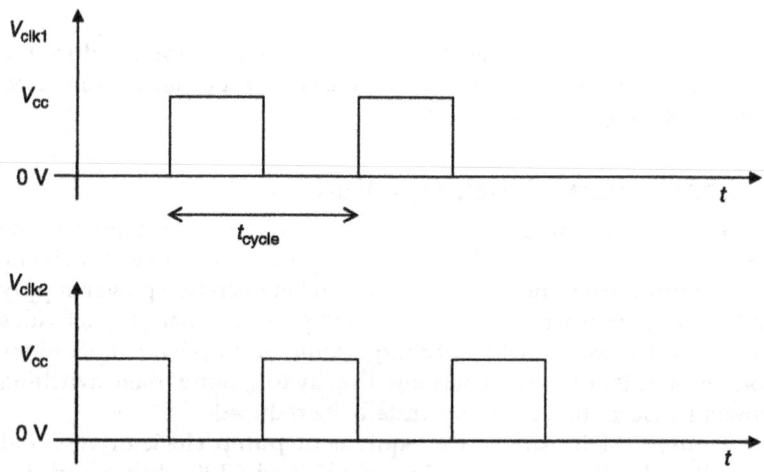

Figure 7.45 Ideal pump clock waveforms.

In the equalization state, final clock drivers are tristated. Charge equalized between CLK_1 and CLK_2 through transistor MN10 to reach an intermediate level. Following equalization the pump clock driver resumes. Through charge recycling, it helps reduce the switching power consumption.

Figure 7.47 illustrates waveforms of pump clock drivers with equalization. As expected, with one clock at V_{cc} and the other one at 0 V, the equalization process moves both clocks to the intermediate level, which is $V_{cc}/2$ in the illustration. After duration of t_{EQ}, the equalization process ends and the pump clock resumes normal operation.

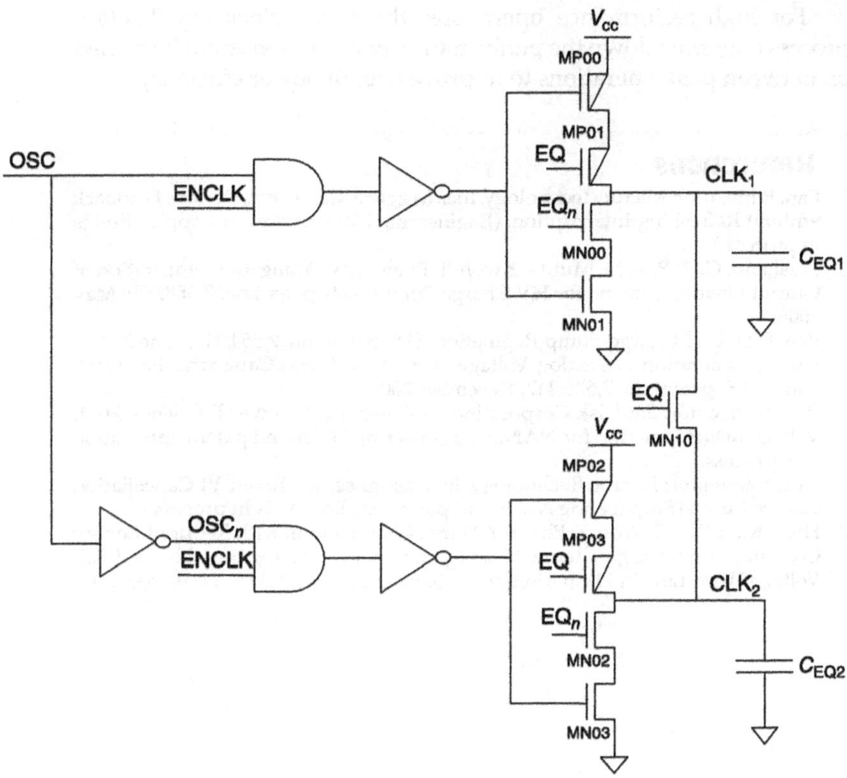

Figure 7.46 Pump clock drivers with charge equalization.

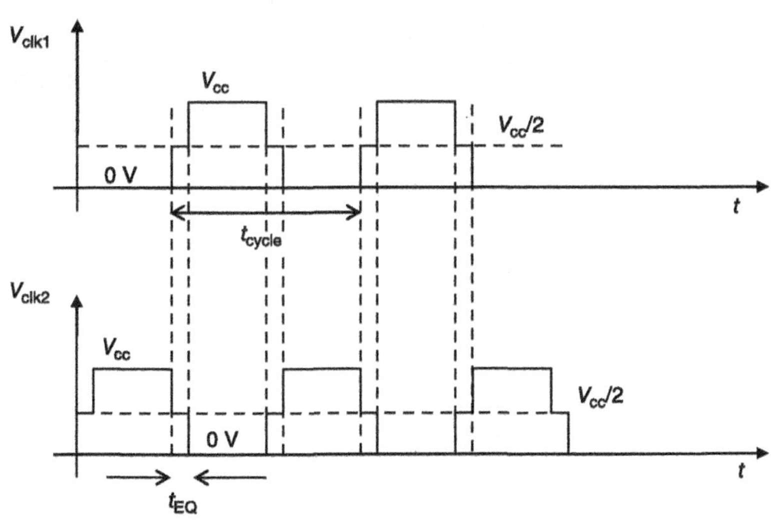

Figure 7.47 Waveforms of pump clock drivers with equalization.

For high-performance operations, the pump clock equalization process may slow down the performance. Equalization could be turned on between peak operations to improve overall power efficiency.

7.12 References

1. Pan, F. inventor; Micron Technology, Inc., assignee. Capacitive Divider Feedback without Refreshing Interruption. (Engineering Notes and patent application is in process.)
2. Prashanti, G., F. Pan, M. Mui, G. Kwon, T. Pham, C.M. Wang. Implementation of Output Floating Scheme for HV Charge Pumps. U.S. patent no. 7,368,979, May 2008.
3. Pan, F. Hybrid Charge Pump Regulation. U.S. patent no. 7,554,311, June 2009.
4. Pan, F. Generation of Analog Voltage Using Self-Biased Capacitive Feedback Stage. U.S. patent no. 7,626,447, December 2009.
5. Pan, F. inventor; SanDisk Corporation, assignee. High Power Efficiency High Voltage Bias Regulation for NAND. (Engineering Notes and patent application is in process.)
6. Pan, F. inventor; Micron Technology, Inc., assignee. RC Based Vt Cancellation Charge Pump (Engineering Notes and patent application is in process.)
7. Htoo, K., F. Pan, B. Woo, T. Pham, Y. Wang. Charge Pump with Reduced Energy Consumption through Charge Sharing and Clock Boosting Suitable for High Voltage Word Line in Flash Memory. U.S. patent no. 8,339,183, December 2012.

CHAPTER 8
Charge Pump Applications

The normal usages of the charge pump are in the domain of power conversion or power conditioning. The charge pump converts input supply voltages to output bias voltages required by design. It could be from low potential voltage to high potential voltage, or from high potential voltage to low potential voltage, or inverting of the polarity of the output supply with respect to input supplies. These applications could be mapped easily into SMPS or linear regulator counterparts.

In this chapter, new applications of charge pump are explored.

When the charge pump is used to convert power, the primary goal is to achieve the highest power conversion efficiency with minimum power loss. This is represented by Eq. (8.1):

$$\text{Power conversion efficiency} = \frac{P_{out}}{P_{total}} = \frac{P_{out}}{P_{out} + P_{quiescent}} \tag{8.1}$$

$$P_{out} = V_{out} I_{out} \tag{8.2}$$

Here P_{out} represents the total output power consumed by loading circuits, and $P_{quiescent}$ represents the additional power consumed by the charge pump that is not delivered to the output load; and P_{out} could be expressed by Eq. (8.2). For DC–DC power conversion, all the discussion so far has been focused on how to minimize $P_{quiescent}$ and how to improve the power conversion efficiency.

What if we looked from a different angle at Eq. (8.1) and expressed the reciprocal of Eq. (8.1) as Eq. (8.3)? The new term is expressed as the power loss amplification ratio (PLAR). After expanding all the terms inside, PLAR can be expressed as Eq. (8.4):

$$\text{Power loss amplification ratio} = \frac{1}{\text{power conversion efficiency}} \tag{8.3}$$

$$\text{PLAR} = \frac{P_{total}}{P_{out}} = \frac{P_{out} + P_{quiescent}}{P_{out}} \tag{8.4}$$

What is the significance of Eq. (8.4)? If $P_{quiescent}$ is assumed to be smaller than P_{out}, then PLAR is approaching 1. From the power conversion point view, this is ultimate goal that design wants to achieve. On the other hand, if P_{out} is much smaller than $P_{quiescent}$ under certain conditions, then Eq. (8.4) could be simplified to Eq. (8.5). There is an amplification effect between power consumed by the charge pump and weak output load.

$$\text{PLAR} \approx \frac{P_{quiescent}}{P_{out}} \qquad (\text{if } P_{out} << P_{quiescent}) \tag{8.5}$$

$$\text{PLAR} \approx \frac{P_{quiescent}}{P_{out}} = \frac{V_{sup} I_{sup}}{V_{out} I_{out}} = K \frac{I_{sup}}{I_{out}} \tag{8.6}$$

Equation (8.6) shows that, under the assumption of $P_{out} << P_{quiescent}$, PLAR could be translated to a constant value multiplied by the ratio of input supply current to output load current. If I_{out} were the signal of interest in the design, then PLAR would be an amplification of signal of interest.

If the signal of interest in this context were too difficult to be measured directly, and/or if the signal in this context were too weak to be detected using normal circuit design, then the amplified signal through the use of the charge pump would be a viable method to detect and quantify the signal of interest.

In analog designs, many proven methods can detect current magnitude under test. Depending on the performance and accuracy requirements, the designer has the flexibility to compare the load current directly or indirectly with a known reference current. Sometimes those methods could still be applied in the domain of high-voltage design. However, sometimes it could be quite challenging to implement those approaches under the given design constraints.

All the following sections are centered on the theme of how to reliably sense and detect the load current under test, and how to use the sensed result and feedback to improve the design in operations. The current of interest could be the load current directly connected to charge pump; or it could be a current of interest connected indirectly through AC coupling.

8.1 Oversampling Charge Pump ADC

The charge pump could be used as an analog-to-digital converter (ADC) to quantify the amplitude of the load current. This is limited to the type of charge pump design using pulse width modulation (PWM), pulse frequency modulation (PFM), or on/off regulation. There is naturally a linear relationship between the output load current and total pulse width (or total pulse counts) within a fixed period in charge pump design. While the output voltage of the charge pump

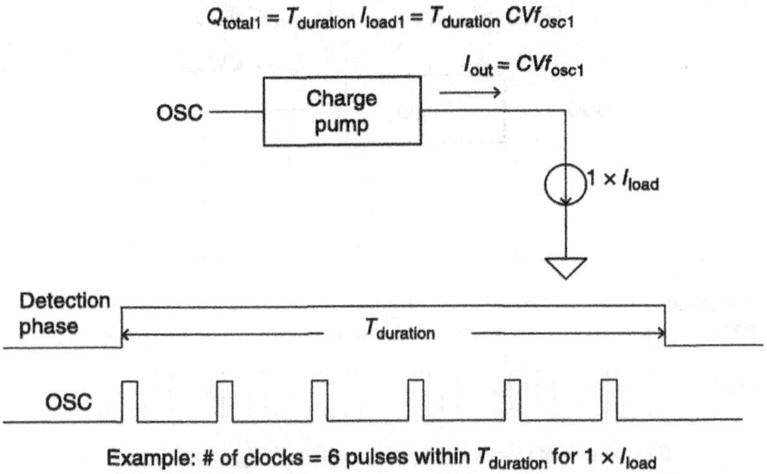

Figure 8.1 Charge pump under regulation with $1 \times I_{load}$.

is being regulated, the conversion between the load current and digital pump clock pulse counts occurs simultaneously and automatically.

Figure 8.1 illustrates the operation of the charge pump under regulation with $1 \times I_{load}$. In steady state under regulation, the average charge pump output current must be equal to $1 \times I_{load}$ current.

$$I_{out} = CVf_{osc} \tag{8.7}$$

Equation (8.7) gives the charge pump output current under regulation where C, V, and f_{osc} are all fixed by design under given equilibrium conditions and would not vary over time. As the example illustrated in Fig. 8.1, for a fixed period $T_{duration}$, a total of 6 pump clock pulses are being recorded within the given duration $T_{duration}$ for $1 \times I_{load}$. The number of pulse counts (PC) within duration $T_{duration}$ could be derived from Eq. (8.8). It is proportional to charge pump oscillation frequency f_{osc}, which is proportional to the load current, as indicated in Eq. (8.7).

$$\text{Pulse count} = T_{duration} f_{osc} \tag{8.8}$$

Figure 8.2 illustrates the charge pump under regulation with $2 \times I_{load}$. As expected, the frequency f_{osc2} of the charge pump must be doubled from f_{osc1}. Since both c and v are fixed terms, with doubling of frequency for $2 \times I_{load}$, within the same $T_{duration}$, there are 12 pump clock pulse counts, as illustrated in Fig. 8.2.

Figures 8.1 and 8.2 illustrate the simple concept of charge pump ADC with respect to current load. There is a linear relationship between the output load current and the switching frequency of the pump clock, as shown in Eq. (8.7), and there is another linear relationship between the pulse counts within a fixed duration, and load current, as shown in

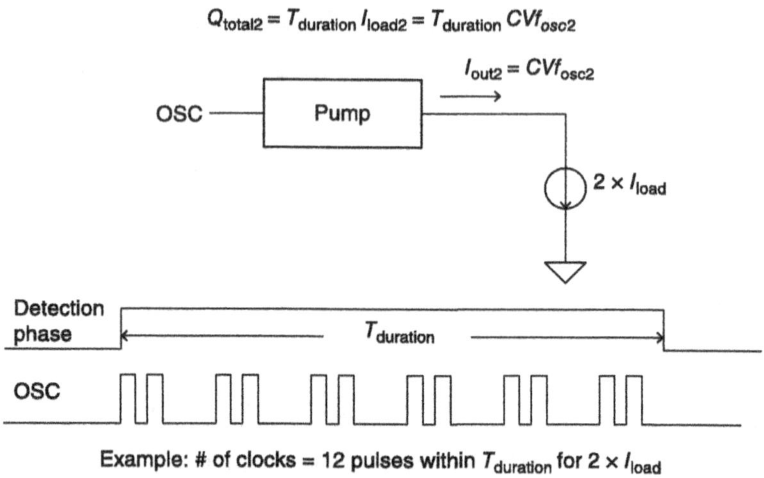

FIGURE 8.2 Charge pump under regulation with $2 \times I_{load}$.

Eq. (8.8). Within a fixed duration of time, the total pump clock pulse count is linearly proportional to the load current. This relationship is illustrated in Fig. 8.3. As the load current doubles, the number of pump clock pulse counts within the fixed duration will double.

If the pulse count within the given sampling time duration could be calibrated with a known reference current, by counting the number of pump clock pulses within the same duration, then the load current could be extrapolated. This is given by Eq. (8.9). Both I_{ref} and PC_{ref} (reference pulse counts for known reference current) are calibrated in an early stage. When the pump pulse count PC is known for a given load, the corresponding current amplitude of load I_{output} could be quantified using Eq. (8.9).

$$I_{out} = \left(\frac{I_{ref}}{PC_{ref}}\right)PC \tag{8.9}$$

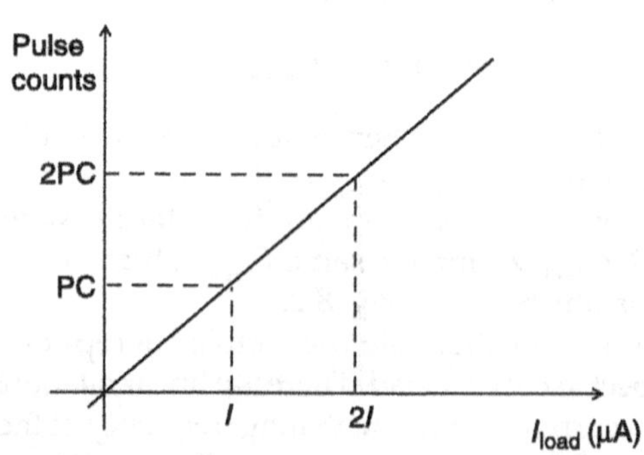

FIGURE 8.3 Pump clock pulse counts versus I_{load}.

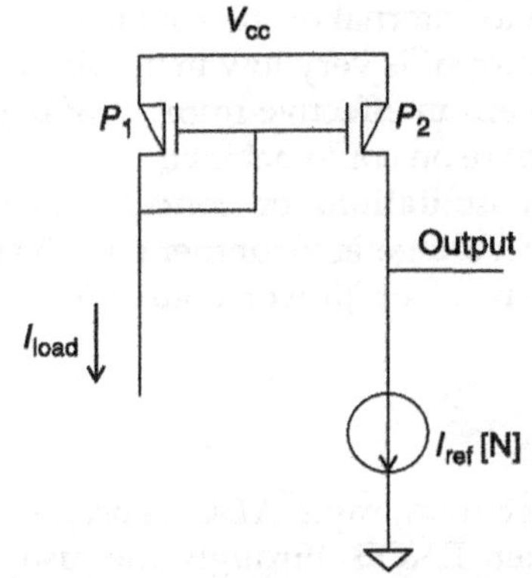

FIGURE 8.4 Current sensing with DC comparison.

This is the foundation of charge pump ADC in order to quantize the load current amplitude under bias. Current could be measured in a few different ways in analog design, such as using DC comparison, or ramp ADC.

8.1.1 DC Comparison

Current sensing could be done with DC comparison between I_{load} and the reference current. Figure 8.4 illustrates the basic implementation of current sensing by DC comparison. Current load is mirrored from P_1 to P_2. The mirrored current from P_2 is compared with I_{ref}[N]. Reference current I_{ref}[N] is swept until the output switches polarity. Special search algorithms could be utilized to speed up the conversion speed.

Figure 8.5 illustrates a simple waveform of DC comparison. Reference current I_{ref}[N] is stepped up until the output switches from V_{cc} to 0 V. Current sensing is a high-speed operation, but the sensed

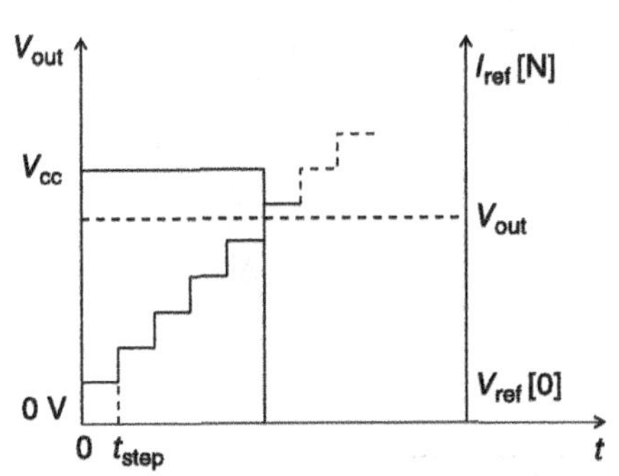

FIGURE 8.5 Waveform of current sensing by DC comparison.

result is susceptible to external or internal noises and mismatches in design. If the load current is very low in amplitude, then there could be an accuracy concern in effective number of bits (ENOB) that this ADC with DC comparison could achieve.

There are other limitations or concerns of implementing DC comparison in high-voltage environment, such as electrical design rules for devices, speed or power consumption beside accuracy concerns.

8.1.2 Ramp ADC

To achieve higher accuracy, ramp ADC[1] is used in current sensing to achieve very higher ENOB through the use of an integrator. Figure 8.6 illustrates the implementation of current sensing by ramp ADC. The load current I_{load} and reference current I_{REF} are mirrored and are integrated over integration capacitance C over different phases of operation. The voltage on the output node is compared with reference voltage V_{REF}. The comparison result is used to control the logic switching to charging or discharging integration capacitance C.

Figure 8.7 illustrates a two-step ramp current sensing scheme. At time zero, load current I_{load} is integrated over capacitance C. The voltage on the top plate of capacitance C is initialized to 0 V. Within the time period t_1, the voltage on node output has reached reference voltage V_{REF}. Logic output compare would turn off switch SW_1 and SW_3 to stop integration of I_{load}, and it would turn on switch SW_2 and SW_4 to start discharge C with reference current I_{REF}. Node output

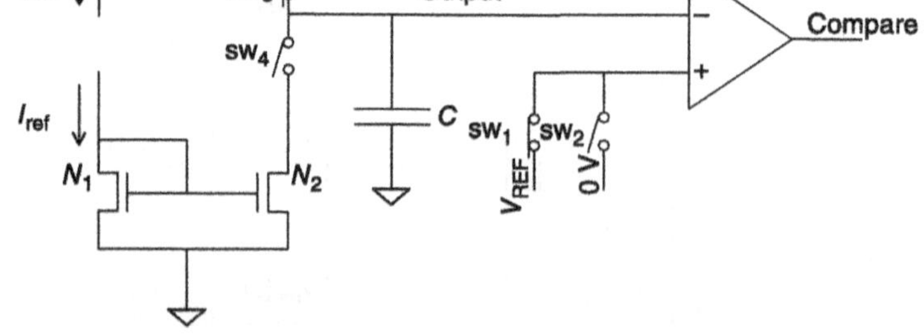

FIGURE 8.6 Current sensing by ramp ADC.

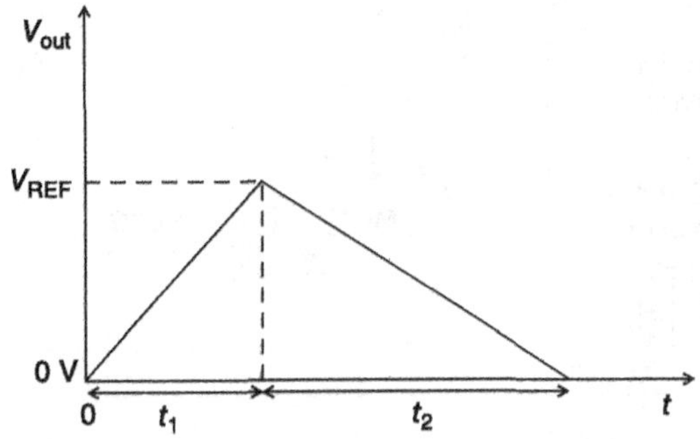

Figure 8.7 Two-step ramp current sensing.

would start discharging from V_{REF}. After duration of t_2, the potential on node output crosses over the 0-V reference and conversion is completed.

The load current could be derived from Eq. (8.10).

$$I_{load} = I_{REF}\frac{t_2}{t_1} \tag{8.10}$$

If t_1 and t_2 could be quantized with the reference clock, and the corresponding pulse counts are PC_1 and PC_2, then the load current could be derived from Eq. (8.11) from I_{REF}, PC_1, and PC_2.

$$I_{load} = I_{REF}\frac{PC_2}{PC_1} \tag{8.11}$$

8.1.3 Oversampling Charge Pump ADC

The concept of oversampling charge pump-based ADC[2] was conceived with the high demand to measure ultralow leakage current under fixed high-voltage bias on the fly with high accuracy and high performance. The conventional techniques would be difficult to implement in such an environment, and either detection speed would be slow or accuracy of detection would be a concern. With the charge pump itself using PWM or PFM for regulation to supply load current, the translation between load current and charge pump operating frequency or pulse counts within a fixed duration occurs automatically on the fly in operations. No additional analog circuits need to be implemented for data conversion. Only a few digital counters plus logic gates are needed to record the digitized values of current in operation and processing of detection results in the digital domain.

Figure 8.8 illustrates an implementation of oversampling charge pump ADC. The only circuit block added to the charge pump system

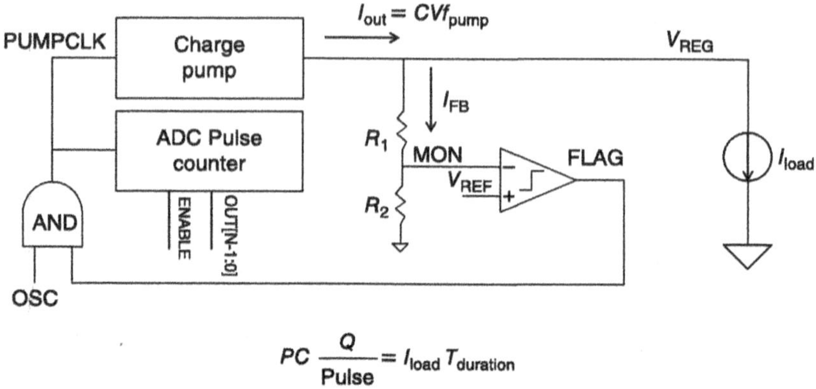

FIGURE 8.8 Oversampling the charge pump ADC.

is the ADC pulse counter with input control signal ENABLE, and digitized output OUT[N-1:0] the

$$I_{load} = I_{output} + I_{FB} \tag{8.12}$$

$$PC\frac{Q}{Pulse} = (I_{load} + I_{FB})T_{duration} \tag{8.13}$$

Equation (8.12) shows the correlation among the load current I_{load}, I_{output}, and I_{FB}. Eq. (8.13) shows that the total charge consumed by I_{load} and I_{FB} within duration $T_{duration}$ should be equal to the product of the total pulse count PC and charge per pulse delivered from the charge pump.

If I_{FB} is negligible compared with I_{load}, then the S/N ratio is high and I_{FB} could be ignored. If I_{FB} is comparable to the signal of interest, then hybrid regulation schemes[3] introduced in Chap. 7 could be adopted in charge pump regulation to cut off DC I_{FB} consumption. This technique would improve the S/N ratio of detection for the charge pump ADC.

Why is the design claimed to be oversampling? In a noisy mixed signal environment, detection and quantization of ultralow-amplitude current would be extremely difficult. Oversampling techniques offer increased resolution to gain accuracy with respect to the oversampling frequency over the signal frequency. For the oversampling charge pump ADC, if the total pulse count representing the amplitude of current within $T_{duration}$ could be increased with respect to decision-making pulse width $T_{duration}$, then the design holds the principle of oversampling ADC.

The tradeoff between performance and accuracy of oversampling charge pump ADC could be easily achieved by changing the sampling time duration $T_{duration}$. The shorter the decision period, the less accurate the quantization result, but the higher the performance that could be achieved. The longer the decision period, the higher the resolution but lower conversion speed would be achieved. Doubling $T_{duration}$ would achieve 2 times more pulse counts, or 6-dB gain in charge pump ADC resolution as outcome of improved signal-to-quantization noise ration (SQNR)[1] with added bit. The ENOB would be determined by proper simulation, and truncation of error bits would be processed in the digital domain.

The oversampling charge pump ADC offers amplification of load current in practice. Equation (8.6) could be converted to Eq. (8.14):

$$\frac{I_{sup}}{I_{out}} = \frac{\text{PLAR}}{K} \tag{8.14}$$

The design of the oversampling charge pump ADC could be optimized to increase the amplification factor between I_{sup} and I_{out}. This is the opposite of the normal approach in charge pump design, in which the goal is to optimize the power conversion efficiency. Increasing the amplification factor of design would allow sensing even lower current amplitude within a given time period, or a faster performance for measurement.

8.2 Sigma-Delta Oversampling Charge Pump ADC

Oversampling charge pump ADC could achieve high resolution in quantization of ultra low current while maintaining high-speed performance. Could this fundamental ADC architecture be further improved to achieve even higher resolution?

Sigma delta and the noise shaping concept are a powerful practice to achieve even higher resolution in ADC design. By shaping the quantization noise in frequency domain, and pushing the noise energy from the low-frequency range into the higher-frequency range that is out of the bandwidth of interest, the final ADC quantization result after decimation filtering would achieve even higher resolution than oversampling ADC.

Figure 8.9 illustrates the concept and implementation of the sigma-delta oversampling charge pump ADC.[4] Compared with Fig. 8.8 of the oversampling charge pump ADC, the newly added portion of circuits related to the sigma-delta charge pump ADC is enclosed by dotted lines. There two transconductance cells Gm1 and Gm2 and one integration capacitance $C_{integrate}$ are added in a new feedback loop back to V_{reg}.

How do sigma-delta and noise shaping operate in the sigma-delta oversampling charge pump ADC, as the example illustrated in Fig. 8.9?

First, the difference in V_{reg} between two consecutive pump clock cycles should be determined. Node MON in the feedback loop samples output voltage V_{reg} continuously. Gm1 is used to convert sampled voltage on node MON to current. In steady state, node MON should fluctuate near V_{REF}, and the output current of Gm1 would ramp up and ramp down around its own equilibrium level. The difference

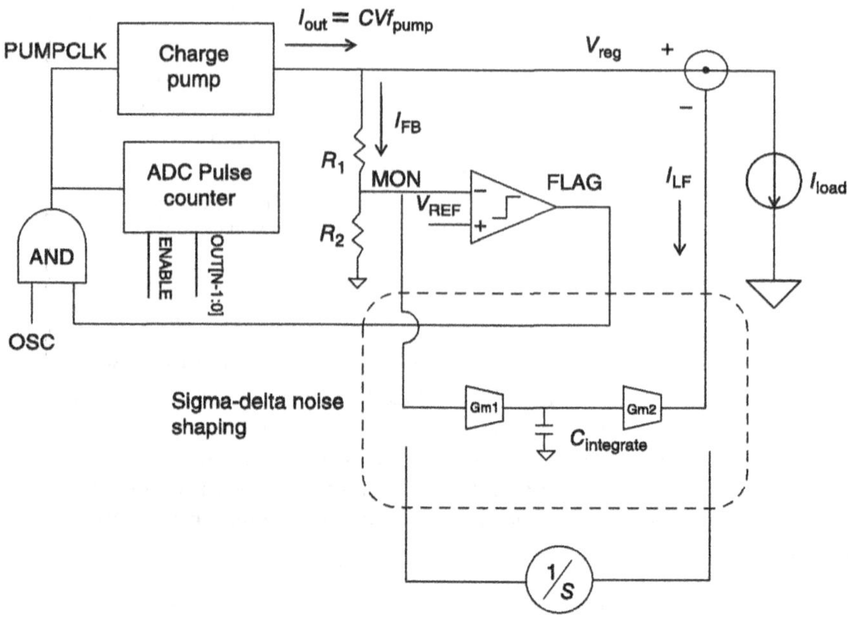

Figure 8.9 Sigma-delta oversampling charge pump ADC.

(delta) of output current over many pump clock cycles would be integrated onto capacitance $C_{integrate}$. The process of integrating output current differences is called *sigma*. In Fig. 8.9, the integration process is highlighted as $1/S$. Integrator $1/S$ attenuates the high-frequency noise and error in the output current. DC and low-frequency noise or error would be maintained as the voltage builds up over $C_{integrate}$.

To implement noise shaping, negative feedback must be integrated into the system in order to remove the low-frequency noise component in the output. This process is achieved through current subtraction. Gm2 is used to convert the low-frequency noise voltage $C_{integrate}$ back into the current domain. Through the final current summation, the low-frequency noise current I_{LF} is subtracted from I_{out}. The difference is equal to I_{load}. Through sigma-delta and noise shaping, the new implementation effectively moves the low-frequency noise energy and pushes it into the high-frequency domain. The final ADC output OUT[N-1:0] would be processed in the digital domain to decimate the high-frequency noise component. Higher resolution or ENOB could be claimed with sigma-delta oversampling charge pump ADC.[1]

Figure 8.10 illustrates the effect of sigma-delta and noise shaping in transient response for oversampling charge pump ADC. In steady state when the charge pump reaches regulation level V_{reg} with load I_{load}, the noise on the output of the charge pump will be large. The waveforms encompass noises in all frequency ranges from load,

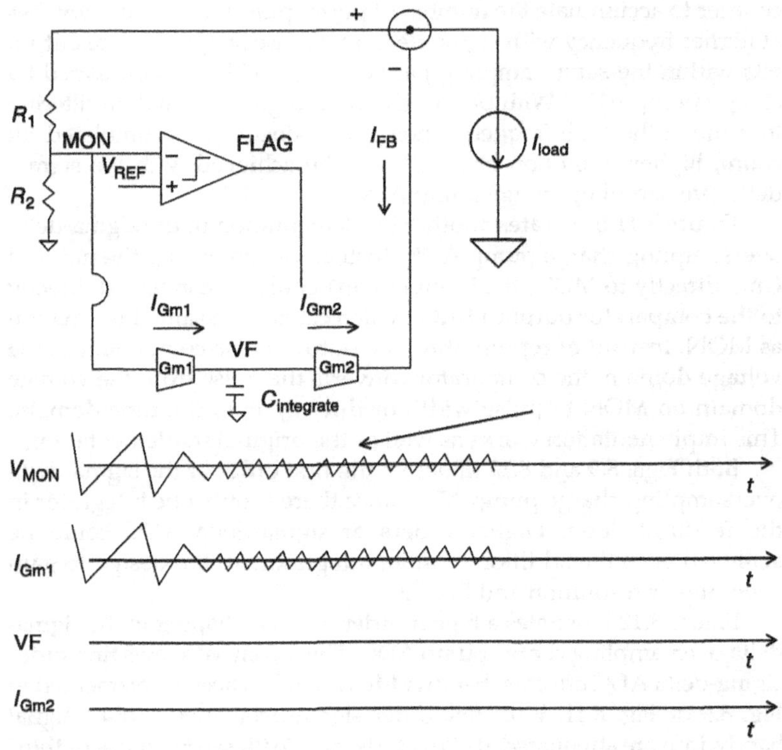

FIGURE 8.10 Noise shaping of sigma-delta oversampling of the charge pump ADC.

supplies, devices, etc. A large portion of noise is located in the low-frequency range. When the feedback network samples the charge pump output, noise is sampled into V_{MON} together with the signal of interest. Gm1 converts information from voltage domain to current domain. The converted current is integrated on $C_{integrate}$. Since integrator $C_{integrate}$ has infinite DC gain and very high low-frequency gain, node VF accumulates the low-frequency portion of the output noise. Higher-frequency noise is attenuated through the integrator. To subtract DC and the low-frequency noise component from the output, which is in terms of current, Gm2 is used to convert VF in the voltage domain back to the current domain. Current summation at the output node subtracts I_{FB} from I_{out}. As time elapses, low-frequency noise is being noise-shaped into the high-frequency domain.

As waveforms of V_{MON} and V_{FB} illustrated in Fig. 8.10, the transient waveforms switch at higher frequency as time elapses. Transient waveforms signal that the frequency of output noise is picking up. Noise in the lower frequency range is being removed over time. With the integrator, node VF and I_{GM2} tend to be stable around equilibrium, and maintain the closed-loop feedback. FLAG triggers the ADC pulse counter to accumulate the number of pump pulses. As FLAG switches at higher frequency with sigma-delta and noise shaping, more output bits within the same sampling period $T_{duration}$ will be accumulated by charge pump ADC. With proper choice of digital decimation filtering to remove the high-frequency noise component on accumulated bit count, higher resolution or ENOB will be achieved with the sigma-delta oversampling charge pump ADC.

Figure 8.11 illustrates another implementation of the sigma-delta oversampling charge pump ADC. Instead of connecting the input of Gm1 directly to MON, the input of Gm1 could be connected directly to the comparator output FLAG, which carries the same information as MON. Instead of representing the output noise component in the voltage domain, the comparator converts the noise from the voltage domain on MON to pulse width or duty cycle in the time domain. This implementation works as well as the original implementation.

Both Figs. 8.9 and 8.11 illustrate the first order of the sigma-delta oversampling charge pump ADC since there is only one integrator in the feedback loop. Higher orders of sigma-delta ADC could be achieved with the addition of more integrators in the design to gain even higher resolution and ENOB.

Figure 8.12 illustrates a higher order of noise shaping in the sigma-delta oversampling charge pump ADC. The top curve shows first-order sigma-delta ADC effective bandwidth, which has been implemented in Fig. 8.9 or Fig. 8.11. With first-order sigma-delta ADC, noises signal bandwidth are attenuated at –20 dB/decade. With second order or third order of noise filtering, noises signal bandwidth would be attenuated at a higher rate, either –40 dB/decade or –60 dB/decade.

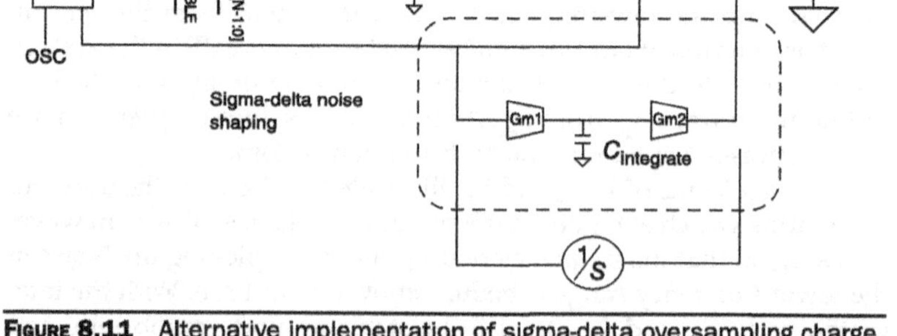

FIGURE 8.11 Alternative implementation of sigma-delta oversampling charge pump ADC.

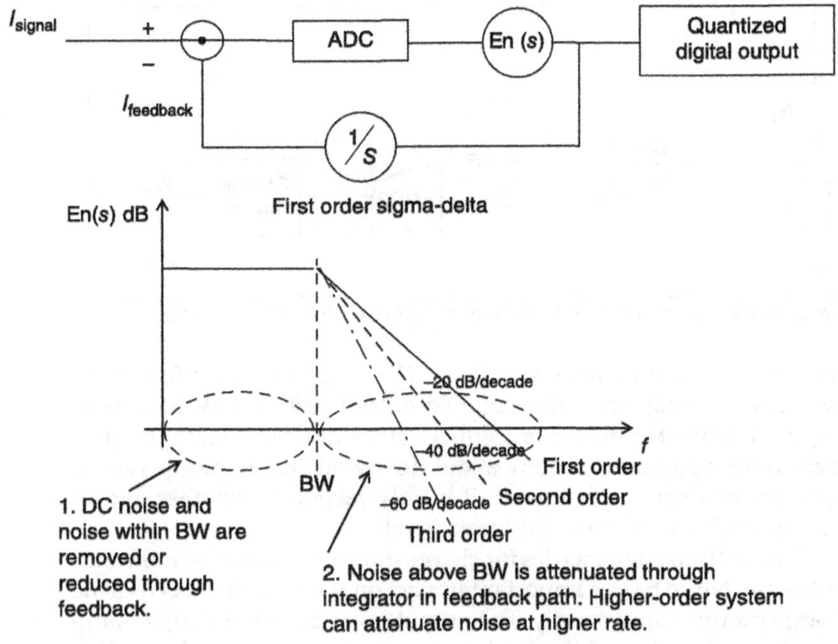

Figure 8.12 Higher order of noise shaping in sigma-delta oversampling charge pump ADC.

With a higher rate of attenuation through higher orders of noise shaping, less noise energy within the signal bandwidth would be included for quantization. Higher resolution and larger ENOB are expected with higher-order sigma-delta oversampling charge pump ADC.

8.3 Dynamic Output Noise Optimization Using Charge Pump-Based Current ADC[5]

With a DC–DC converter, such as a charge pump design, it is optimized in design phase for performance, noise, and power conversion efficiency under particular load condition. The condition is normally set by the absolute worst cases in loading and PVT corners.

However, in real usage the loading condition could change dynamically. PVT corners could vary depending on the process corner, temperature, and supply voltages. The reality makes the charge pump design a challenge in the field of operations. Either overdesign should be done in the design phase to take extra margins, or in operations it could cause the performance to deviate from the target.

Negative feedback is commonly used in analog or digital designs to reduce variations that could not be anticipated. In charge pump architecture, the output voltage is controlled and regulated with

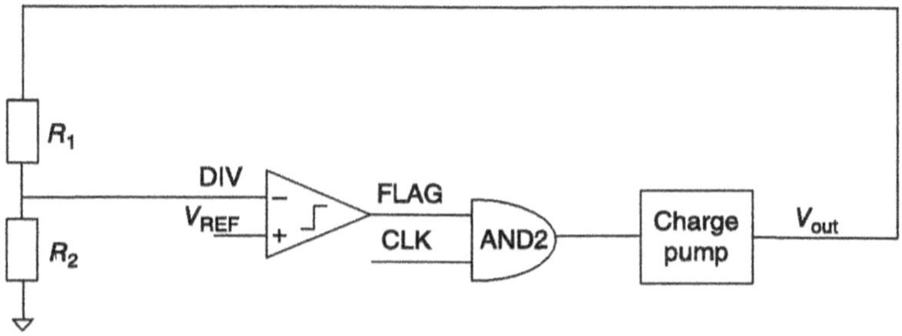

Figure 8.13 Normal negative feedback control loop of charge pump.

respect to a known reference voltage. The only feedback information available to designers is the sampled output voltage DIV, as seen in Fig. 8.13. Unfortunately this feedback does not carry enough information for designers to know about how well charge pump system operates in a real environment. It is only related to whether output voltage is above or below the target level.

One of the design metrics for charge pump performance is the output noise. Noise has an important impact on the load that receives the charge pump output supply. The output noise from the charge pump is a strong function of the load current, output current, decoupling capacitance, and bandwidth of feedback path. If only the amplitude of output voltage were available, it would be impossible to reconfigure the charge pump design dynamically to minimize the output noise in the real environment. This puts a lot of pressure on the charge pump design if loading or PVT conditions could vary over a wide range.

How can one dynamically scale the charge pump designs to minimize output noise in the field of operations? Additional information besides the amplitude of the output voltage must be fed back to circuits in real time in order to dynamically reconfigure or reoptimize the charge pump system with respect to the change of load or PVT corners.

Output supply noise is determined by the power delivered from the charge pump, power consumed by the load, and total decoupling capacitance. The difference between output current and load current in regulation is integrated by the decoupling capacitance on the output of the charge pump. With output voltage being regulated to target, how is one to extract the load current information? With the charge pump ADC introduced earlier, load current could be extracted dynamically in operation. There is no need to insert any series circuits in the path where the output current flows. If the load current is known at a given time, how could the design be optimized dynamically?

Figure 8.14 illustrates an example of pump clock pulse counts within a 1-μs sampling window for different I_{load}. Both pulse1 and pulse2 are sampling windows with 1-μs duration. They are alternating in operation. As illustrated in the figure, within pulse1, there are four different cases for pump with four different loading currents.

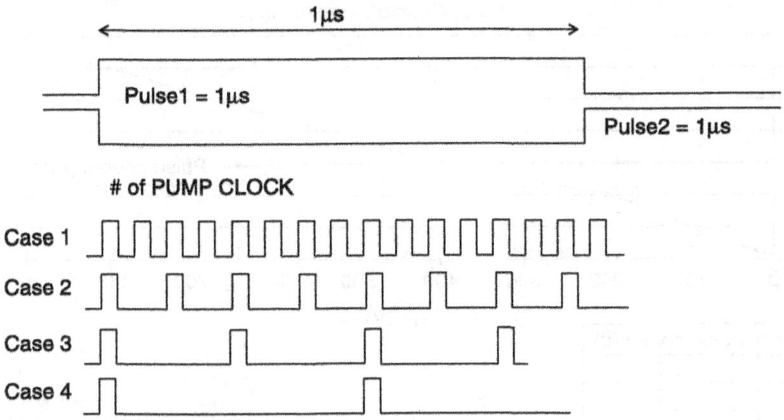

FIGURE 8.14 Pump clock pulse counts within 1-μs sampling window for different I_{load}.

Case 1 has 15 pulse counts in the 1-μs window. Case 2 has 8 pulses. Case 3 has 4 pulses and case 4 has 2 pulses. Based on the charge pump ADC output, case 4 has the least amount of current load, while case 1 has the largest amount of current load. From the feedback of the magnitude of the load current some procedures could be applied to minimize the output noise if it had higher priority in design.

The following approach is a design example of how to reconfigure the charge pump dynamically to minimize output noise with the knowledge of the magnitude of load current in regulation.

Figure 8.15 illustrates an example of the trend line of pump clock pulse counts versus I_{load} in a 1-μs sampling window. The solid line is generated from simulation results. There is nonlinearity in design and simulation. However, the dotted trend line clearly illustrates a linear relationship between pulse counts in 1-μs sampling window and load current as expected.

Figure 8.16 illustrates the trend line of the charge pump output noise versus I_{load} with the same test case as in Fig. 8.15. The solid line is generated from real data points from simulation results. The dotted line shows a linear trend line. There is nonlinearity at the low output current load range. However, in general as load current increases, the linear relationship between output noise and I_{load} is apparent.

The linear relationship between output noise and output load current could be easily derived from the following equations based on simple assumptions.

$$I_{diff} = I_{out} - I_{load} \tag{8.15}$$

$$\Delta V = \frac{\int_0^{T_{cycle}} I_{diff}\,dt}{C_{load}} = K - \frac{\int_0^{T_{cycle}} I_{load}\,dt}{C_{load}} = K - \frac{T_{cycle}}{C_{load}} I_{load} \tag{8.16}$$

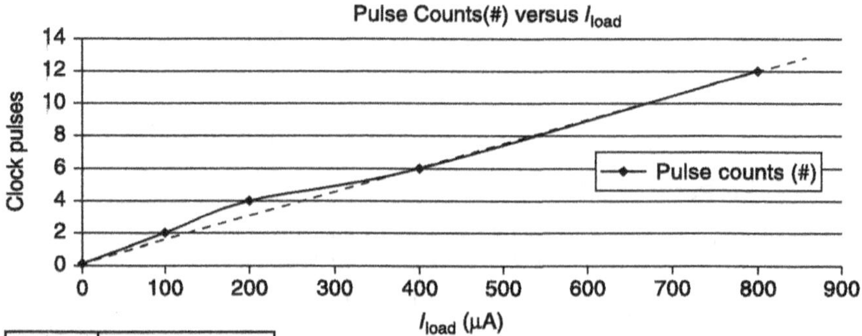

I_{load} (μA)	Pulse counts(#)
0	0.12
100	2
200	4
400	6
800	12

Figure 8.15 Example: trend line of pump clock pulse counts versus I_{load} in 1-μs in sampling window.

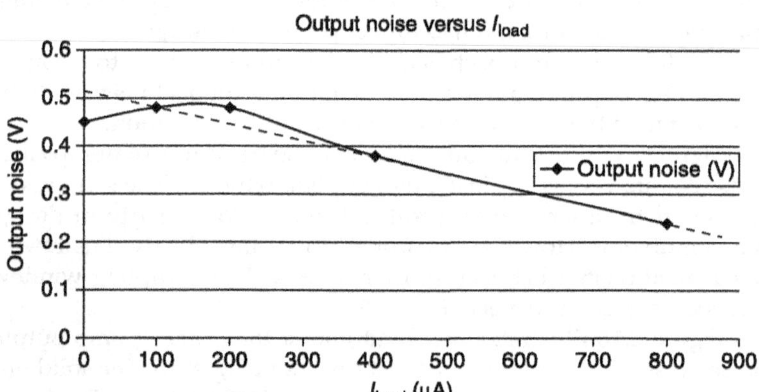

I_{load} (μA)	Output noise (V)
0	0.45
100	0.48
200	0.48
400	0.38
800	0.24

Figure 8.16 Example: trend line of charge pump output noise versus I_{load}.

The noise on the charge pump output node is caused by the difference between load power consumption and charge pump output power. Since the voltage on the output node does not vary in large scale, the difference in power could be viewed as a difference in charge pump output current and load current. This is illustrated in Eq. (8.15). If I_{diff} is integrated over one clock cycle with respect to the load capacitance C_{load}, the output noise ΔV could be derived. Equation (8.16) illustrates the derivation of output noise. In steady state, I_{out} from the charge pump is constant. The load current is the variable. Equation (8.16) shows a linear relationship between output noise and I_{load}.

From Figs. 8.15 and 8.16, the experimental data could be plotted: output noise versus load current, as illustrated in Fig. 8.17. The trend line illustrates a clear linear relationship trend between charge pump output noise and pulse counts within a 1-μs sampling window. The data projects with higher pump clock pulse counts within a fixed time duration, then charge pump output noise would be minimized.

Figure 8.17 would help the designer to develop an algorithm that can optimize charge pump output noise dynamically in real operation regardless of the variation of I_{load} and PVT corners.

Figure 8.18 illustrates an example of a charge pump output noise reduction algorithm. Figure 8.19 illustrates an example of how charge pump output noise is minimized dynamically over four consecutive 1-μs windows.

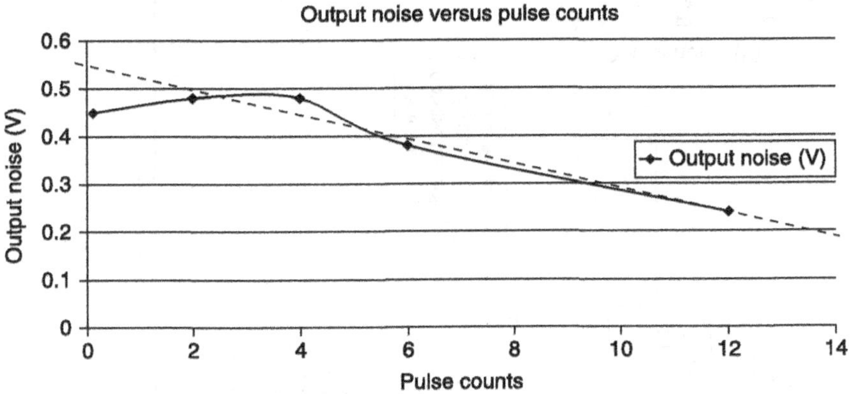

Pulse counts(#)	Output noise (V)
0.12	0.45
2	0.48
4	0.48
6	0.38
12	0.24

FIGURE 8.17 Trend line of charge pump output noise versus pulse counts within 1-μs window.

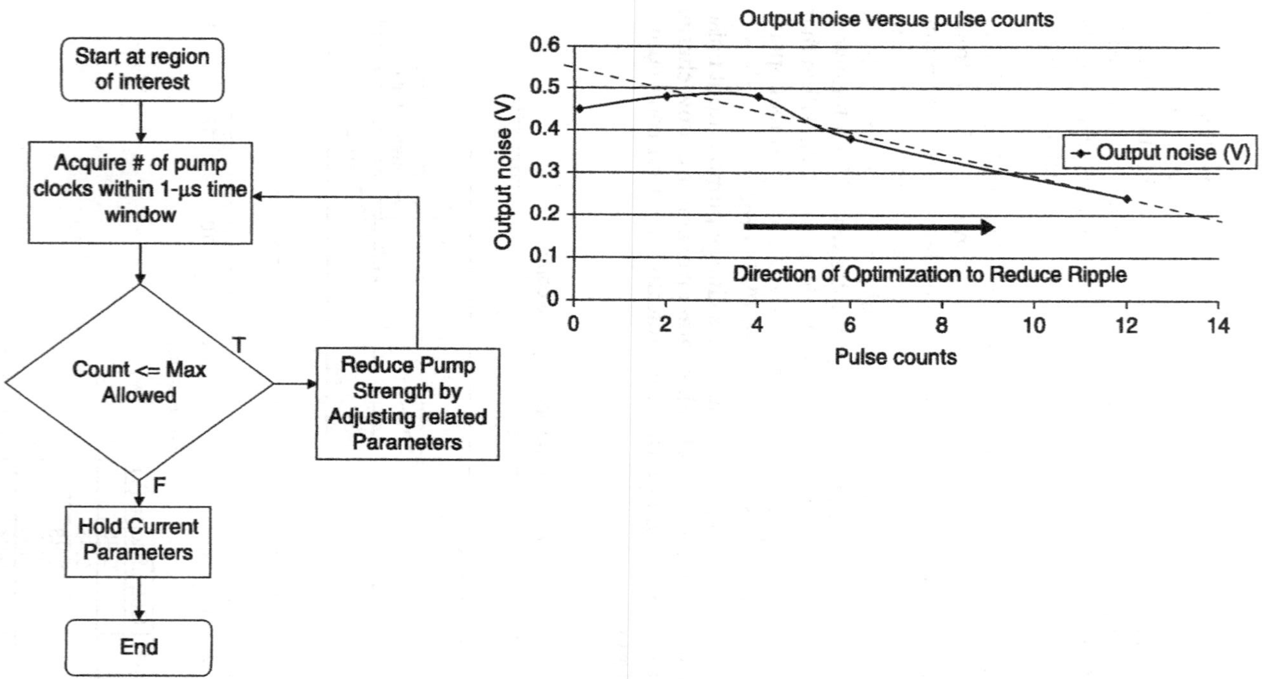

Figure 8.18 Output noise reduction algorithm.

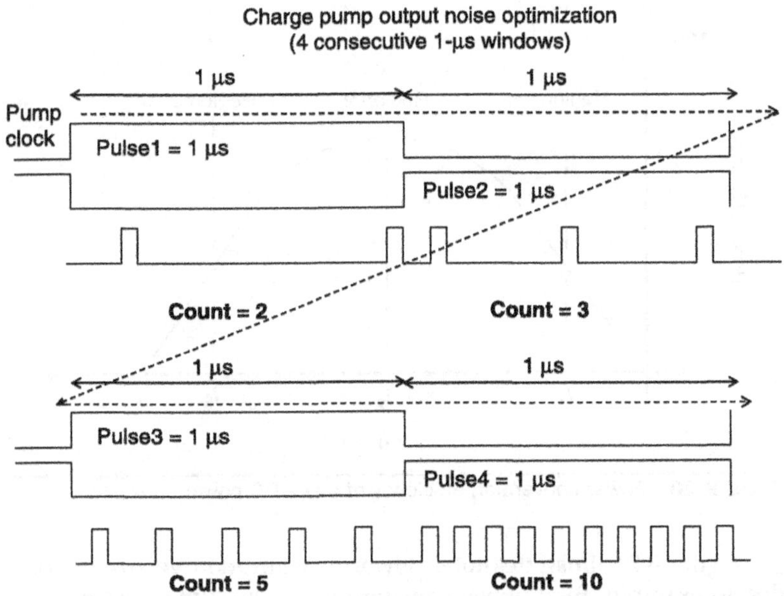

Figure 8.19 Example of dynamical output noise optimization.

8.4 Dynamic Power Efficiency Optimization Using Charge Pump ADC[5]

Power efficiency is another important design metric for charge pump design.

Figure 8.20 illustrates the power efficiency of a DC–DC converter as a function of I_{load}. For a fixed configuration, the optimum power efficiency is in region 2. Whenever the load moves away from region 2 into either region 1 or region 3, the power efficiency degrades from the optimum level. If the charge pump could be reconfigured dynamically based upon the load current, then there would be an opportunity to achieve optimum power efficiency dynamically under any loading condition.

To optimize the power efficiency of charge pump requires the feedback of average DC load dynamically. Ordinary current sampling and feedback would be either too expense or impractical. Charge pump based ADC gives designers the opportunities to extract the average DC current load information without disturbing the original design, and the feedback is achieved dynamically in operation.

In the following discussion the correlation between the power consumption and output noise would show the procedures to optimize the power conversion efficiency of the charge pump.

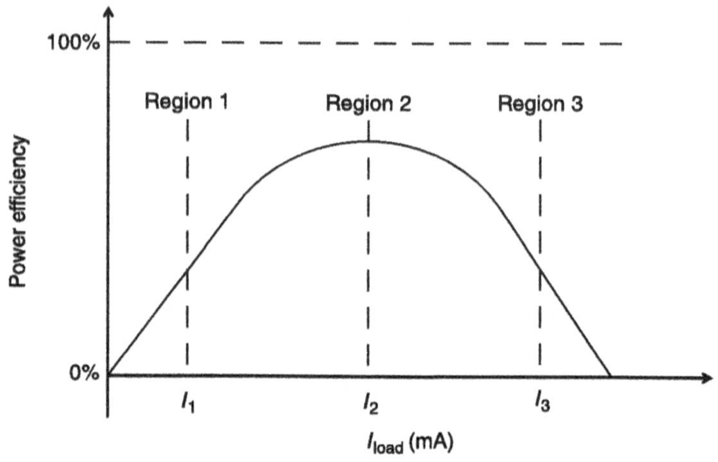

FIGURE 8.20 Power conversion efficiency of a DC–DC power converter.

Figure 8.21 illustrates total power consumption versus I_{load} of a design example. Total power consumption is the sum of output load power and quiescent power. The trend line shows a linear relationship between total power consumption and I_{load}. As I_{load} increases, the

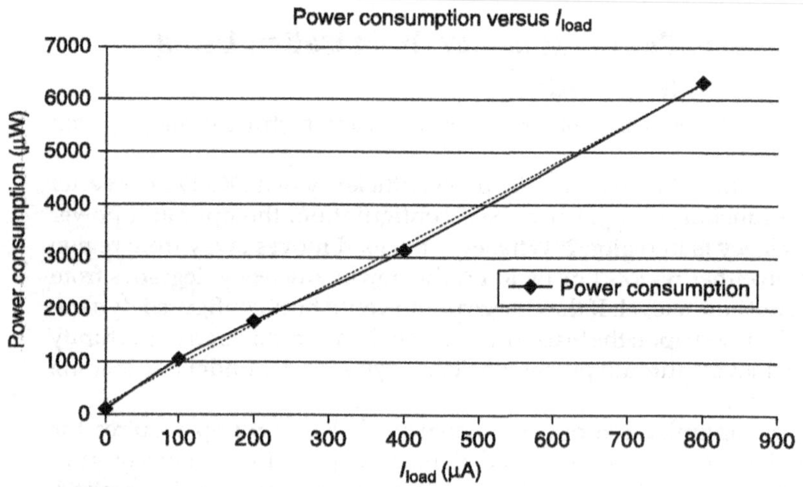

I_{load} (μA)	Power consumption (μW)
0	100
100	1050
200	1770
400	3125
800	6325

FIGURE 8.21 Example: charge pump power consumption versus I_{load}.

power consumption increases. The quiescent power is composed of two components. The first component is a fixed value due to bias and reference used in op-amp or comparator. The second component is related to the switching power consumption of clock drivers, as shown in Eq. (8.17).

$$P_{switching} \propto C_{clk} V_{sup}^2 f \tag{8.17}$$

$$f \propto I_{load} \tag{8.18}$$

Here C_{clk} is the total capacitance being switched in operation, and V_{sup} is the supply for clocks and clock buffers. The supply could be a function of chip input supply; f is the switching frequency of the pump clock, which is proportional to I_{load}, as shown in Eq. (8.18). As I_{load} increases, the quiescent power related to the clock switching will increase as Eq. (8.17) indicates. It will have dominant weight in quiescent power consumption as I_{load} increases. That is why the power consumption has a first-order linear relationship between power consumption and I_{load}.

Using the same design, the output noise of charge pump versus loading current is plotted in Fig. 8.22. Ignoring the nonlinearity at very low-load current, there is a first-order linear relationship between charge pump output noise and I_{load}. As I_{load} increases, the amplitude of output noise reduces linearly. The details are discussed in Sec. 8.3.

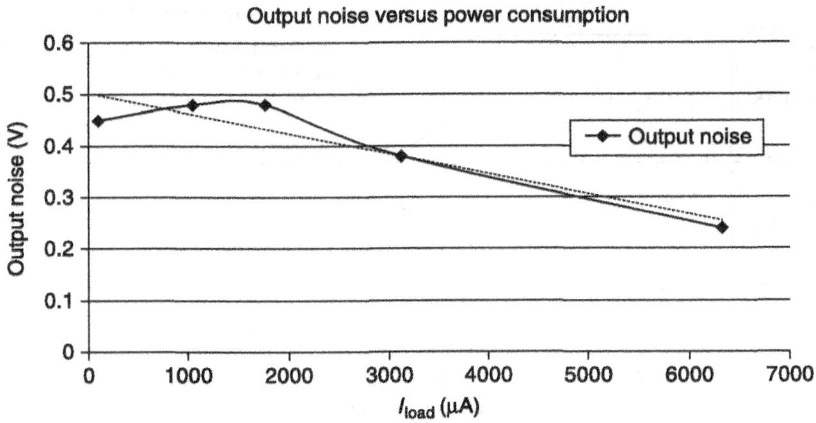

Power consumption (μW)	Output noise (V)
100	0.45
1050	0.48
1770	0.48
3125	0.38
6325	0.24

FIGURE 8.22 Output noise versus power consumption.

One interesting relationship between power consumption and output noise could be found in Fig. 8.23. The greater the output noise that the design can tolerate, the lesser power consumption the charge pump circuit would have. The design could be dynamically configured to optimize between power consumption and output noise as long as both are within the design specifications.

Finally Fig. 8.24 illustrates the relationship between charge pump power consumption and pump pulse counts within a 1-μs window. The solid line is based upon the simulation result, while the dotted line is the trend line that demonstrates the linear relationship between charge pump power consumption and pump clock pulse counts, within 1 μs sampling window.

In normal charge pump design, only the output voltage is sampled for regulation. With the introduction of the charge pump based ADC, output current could be sampled and additional feedback control could be used to optimize the power consumption of the charge pump dynamically.

Figure 8.25 illustrates a simple algorithm to optimize charge pump power consumption in operation. Within the given limits, as long as

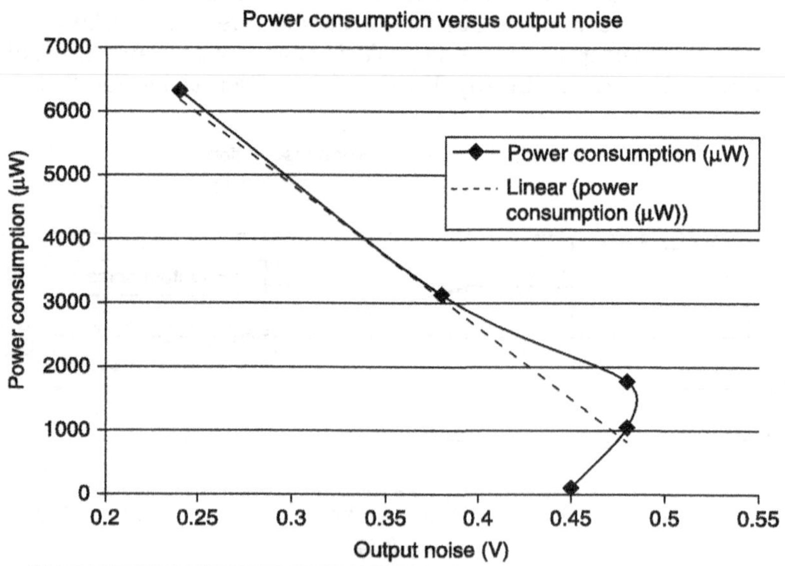

Output Noise (V)	Power consumption (μW)
0.45	100
0.48	1050
0.48	1770
0.38	3125
0.24	6325

FIGURE 8.23 Power consumption versus output noise.

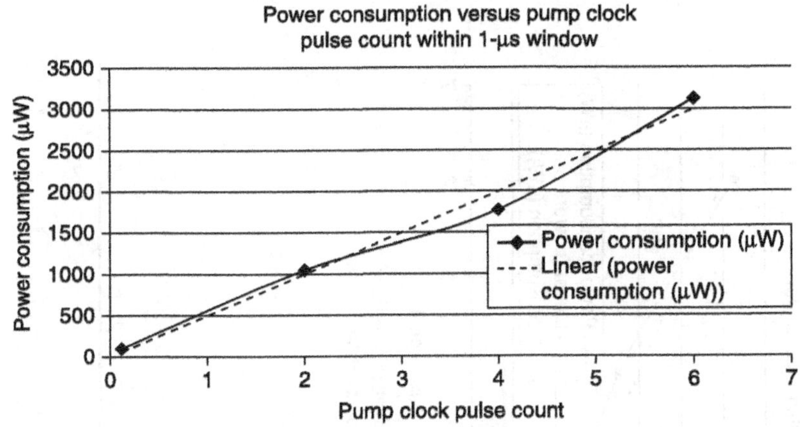

FIGURE 8.24 Charge pump power consumption versus pump clock pulse count within 1-μs window.

output noise is within the acceptable range, fewer switching events would lead to less power consumption in operation.

Figure 8.26 illustrates an example of dynamic power consumption optimization for the charge pump. The pump clock pulse counts are accumulated within a 1-μs sampling window. As long as pulse counts have not reached the target, the charge pump is reconfigured to reduce power consumption. Eventually after 3 μs pulse counts have been reduced from 10 to 2 pulse counts per 1 μs window so is the reduction of switching power.

8.5 Dynamic *IR* Drop Compensation with Variable Current Load

Another usage of charge pump ADC could be dynamic *IR* drop compensation[6] with variable load current. If the charge pump were used to bias a resistive load through a high-resistive path, it would be difficult to determine the exact bias at the far end where the resistive load reside. Feedback is taken near the charge pump output to ensure the target regulation right at the charge pump output and also stability reason. Far end bias near resistive load was not in the control loop and could deviate from target regulation level by *IR* drop.

With the load current information available, the regulation level of the charge pump could be changed dynamically at charge pump output to compensate the *IR* drop along the resistive path to ensure the bias near resistive load is on target. With charge pump based ADC, load current information could be easily obtained for dynamic *IR* drop compensation purposes. Although the accuracy of dynamic *IR* drop compensation would not be the same as sampling the voltage

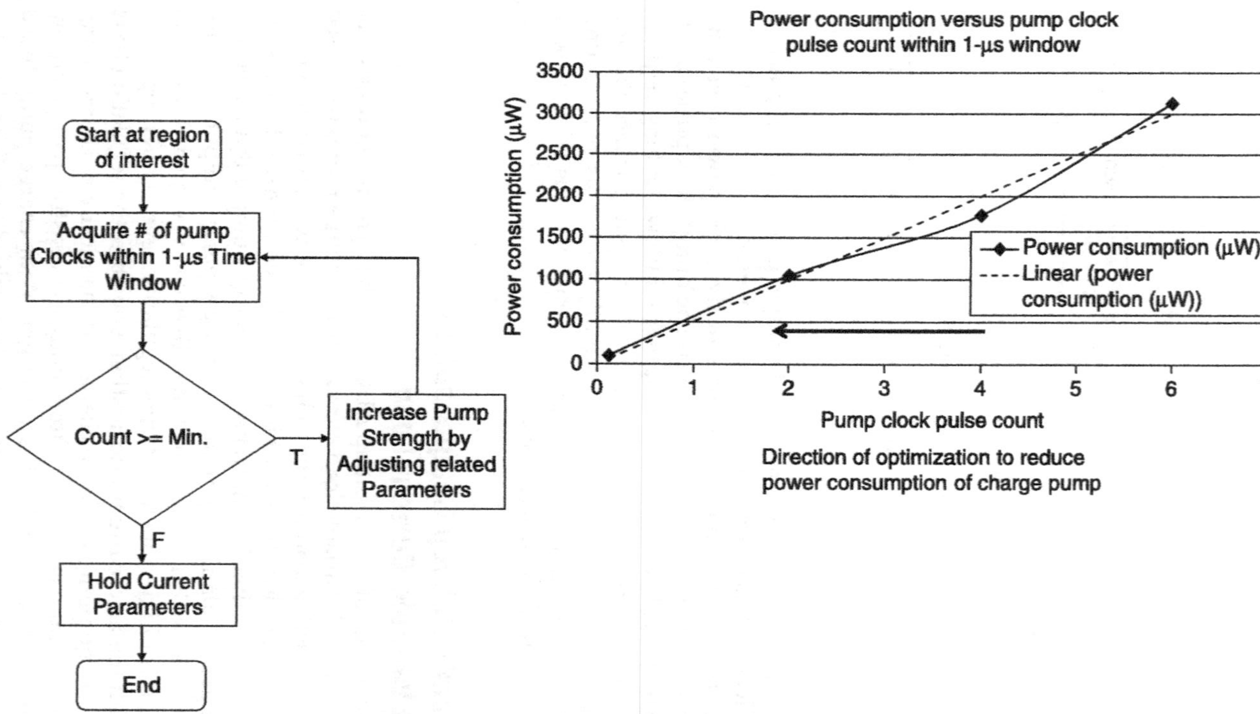

FIGURE 8.25 Charge pump power consumption reduction algorithm.

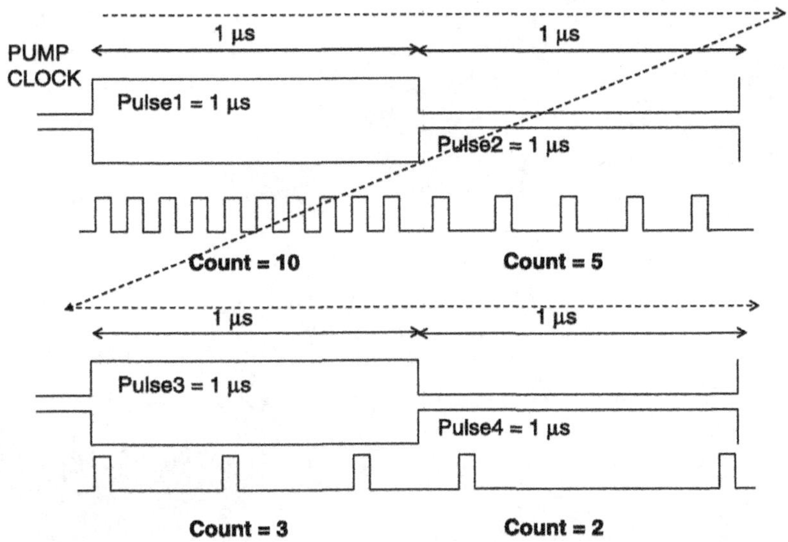

FIGURE 8.26 Example of dynamic power consumption optimization for charge pump.

directly near the resistive load if it could be implemented, with the constraints of layout or technical limitations, it is probably the most cost-effective way to allow dynamic compensation of *IR* drop in those designs.

8.6 References

1. Carusone, T.C., D. Jonhs, K. Martin. *Analog integrated Circuit Design*, 2nd ed., John Wiley, Hoboken, NJ, 2011.
2. Pan, F., J. Wan, S. Guhados, B. Lei, inventors; SanDisk Corporation, assignee. Charge Pump Based Over-Sampling ADC for current Detection. United States patent application US20140084936, A1; 2014 March 27.
3. Pan, F. Hybrid Charge Pump Regulation. U.S. patent no. 7,554,311, June 2009.
4. Pan, F., S. Guhados, inventors; SanDisk Corporation, assignee. Sigma Delta Over-Sampling Charge Pump Analog-To-Digital Convertor. United States patent application US20140085985, A1; 2014 March 27.
5. Pan, F. Dynamic Scaling of Charge Pump Performance Metrics through Current to Pulse ADC Convertor. (Patent application is in process.)
6. Pan, F. Dynamic IR Drop Compensation with Variable Current Load. (Engineering Notes. Patent application is in process.)

Index

Note: Page numbers followed by *f* denote figures.

E

F

G

H

S